MW01436121

Current Topics in Microbiology and Immunology

Volume 388

Series editors

Rafi Ahmed
School of Medicine, Rollins Research Center, Emory University, Room G211, 1510 Clifton Road, Atlanta, GA 30322, USA

Klaus Aktories
Medizinische Fakultät, Institut für Experimentelle und Klinische Pharmakologie und Toxikologie, Abt. I, Albert-Ludwigs-Universität Freiburg, Albertstr. 25, 79104 Freiburg, Germany

Richard W. Compans
Department of Microbiology and Immunology, Emory University, 1518 Clifton Road, CNR 5005, Atlanta, GA 30322, USA

Max D. Cooper
Department of Pathology and Laboratory Medicine, Georgia Research Alliance, Emory University, 1462 Clifton Road, Atlanta, GA 30322, USA

Jorge E. Galan
Boyer Ctr. for Molecular Medicine, School of Medicine, Yale Universitym, 295 Congress Avenue, room 343, New Haven, CT 06536-0812, USA

Yuri Y. Gleba
ICON Genetics AG, Biozentrum Halle, Weinbergweg 22, 06120 Halle, Germany

Tasuku Honjo
Faculty of Medicine, Department of Medical Chemistry, Kyoto University, Sakyo-ku, Yoshida Kyoto 606-8501, Japan

Yoshihiro Kawaoka
Influenza Research Institute, University of Wisconsin-Madison, 575 Science Drive, Madison, WI 53711, USA

Bernard Malissen
Centre d'Immunologie de Marseille-Luminy, Parc Scientifique de Luminy, Case 906, 13288, Marseille Cedex 9, France

Michael B.A. Oldstone
Department of Immunology and Microbial Science, The Scripps Research Institute, 10550 North Torrey Pines Road, La Jolla, CA 92037, USA

Rino Rappuoli
Novartis Vaccines, Via Fiorentina 1, Siena 53100, Italy

Peter K. Vogt
Department of Molecular and Experimental Medicine, The Scripps Research Institute, 10550 North Torrey Pines Road, BCC-239, La Jolla, CA 92037, USA

Honorary Editor: Hilary Koprowski (deceased)
Formerly at Biotechnology Foundation, Inc., Ardmore, PA, USA

More information about this series at http://www.springer.com/series/82

Juan J. Lafaille · Maria A. Curotto de Lafaille
Editors

IgE Antibodies: Generation and Function

Responsible Series Editor: Tasuku Honjo

Springer

Editors
Juan J. Lafaille
Skirball Institute and Department
 of Pathology
New York University School of Medicine
New York
USA

Maria A. Curotto de Lafaille
Singapore Immunology Network (SIgN)
Agency for Science, Technology and
 Research
Singapore
Singapore

ISSN 0070-217X ISSN 2196-9965 (electronic)
Current Topics in Microbiology and Immunology
ISBN 978-3-319-13724-7 ISBN 978-3-319-13725-4 (eBook)
DOI 10.1007/978-3-319-13725-4

Library of Congress Control Number: 2014956362

Springer Cham Heidelberg New York Dordrecht London
© Springer International Publishing Switzerland 2015
This work is subject to copyright. All rights are reserved by the Publisher, whether the whole or part of the material is concerned, specifically the rights of translation, reprinting, reuse of illustrations, recitation, broadcasting, reproduction on microfilms or in any other physical way, and transmission or information storage and retrieval, electronic adaptation, computer software, or by similar or dissimilar methodology now known or hereafter developed.
The use of general descriptive names, registered names, trademarks, service marks, etc. in this publication does not imply, even in the absence of a specific statement, that such names are exempt from the relevant protective laws and regulations and therefore free for general use.
The publisher, the authors and the editors are safe to assume that the advice and information in this book are believed to be true and accurate at the date of publication. Neither the publisher nor the authors or the editors give a warranty, express or implied, with respect to the material contained herein or for any errors or omissions that may have been made.

Printed on acid-free paper

Springer International Publishing AG Switzerland is part of Springer Science+Business Media (www.springer.com)

Preface

IgE antibodies were discovered in the 1960s as "reaginic antibodies" (Ishizaka and Ishizaka 1967), defined at that time as a type of immunoglobulin that could transfer hypersensitivity to protein antigens. Since then, the genes and proteins of both IgE and its high-affinity receptor, FcεRI (Kinet 1999), have been fully characterized. The role of IgE in human chronic inflammatory allergic diseases was suggested by a large number of association studies and was more recently conclusively proven by the beneficial effects of anti-IgE therapy in severe asthma.

The study of IgE-producing B cells was nevertheless long hampered by their relative paucity in peripheral blood and lymphoid organs. The recent development of genetically modified mouse models to study IgE responses has enabled the tracking and isolation of IgE-expressing cells and has uncovered distinctive aspects of the IgE response. We now know that IgE cell responses are characterized by a transient IgE germinal center phase, a predominance of the IgE plasma cell fate, an absence of functional IgE memory B cells, and the co-opting of the sequential switching process to generate high-affinity IgE antibodies. These findings have suggested a link between sequential switching to IgE and the pathogenic potential of IgE responses. Recent advances describing unique features of IgE cell differentiation and of the mechanisms involved in the memory of IgE responses are discussed in this volume by He Jin-shu and collaborators ("Biology of IgE Production: Ige Cell Differentiation and the Memory of Ige Responses").

IgE antibody genes are generated through class switch recombination (CSR), a process whereby the DNA sequence encoding the constant region of the IgE heavy chain, Cε, is brought in close proximity to the VDJ gene through looping out, ligation, and excision of the intervening DNA sequences. Antibody genes can be further modified by somatic hypermutation (SH). Both CSR and SH are mediated by the enzyme AID (Muramatsu et al. 2000). Here Pei Tong and Duane Vesemann discuss the molecular mechanisms underlying the generation of antibodies, how the specific features of CSR to IgE regulate the production of IgE antibodies in immature and mature B cells, and the implications of this regulation for the roles of IgE in disease ("Molecular Mechanisms of IgE Class Switch Recombination").

The pathogenic potential of IgE antibodies lies mainly in its ability to bind to the high-affinity FcεRI on the surface of mast cells and to induce mast cell degranulation upon crosslinking by allergens. Accordingly, it is now well established that IgE is a valid therapeutic target for difficult-to-treat allergic inflammatory diseases, such as corticosteroid-resistant asthma. Stephanie Logsdon and Hans Oettgen ("Anti-IgE Therapy: Clinical Utility and Mechanistic Insights") review the latest clinical data on the treatment of asthma and other allergic diseases with Omalizumab, the first therapeutic anti-IgE antibody successfully used in chronic allergic diseases. Immunological studies in anti-IgE treated patients have also shed new light on the role of IgE in immune responses beyond mast cell activation, and the success of Omalizumab has reinvigorated the search for new approaches targeting IgE-FcεRI interactions.

The affinity of IgE for an allergen, the relative concentration of allergen-specific IgE in the total IgE pool, and IgE cross-reactivity, are all important determinants of the occurrence, or not, of allergic reactions. Ryo Suzuki and collaborators ("New Insights on the Signaling and Function of the High Affinity Receptor for IgE") review the mechanisms of IgE- and FcεRI-mediated mast cell responses, sharing new insights into their modulation: we now know that monomeric IgE can induce diverse signaling and activation responses in mast cells, even in the absence of cognate antigens. New concepts have provided a framework linking monomeric IgE polyreactivity and its ability to activate mast cells in the absence of known antigen; furthermore, the affinity of IgE has been identified as a key determinant in the type of signaling and inflammatory responses induced downstream of antigen recognition.

The pathogenic role of IgE in allergy has somehow eclipsed its valuable protective functions in the case of parasitic infection. Helminthes in particular induce the production of high levels of serum IgE, but their presence is, somewhat counterintuitively, associated with protection from allergic diseases, even in the presence of IgE antibodies against allergens. Firdaus Hamid and collaborators ("Helminth-Induced IgE and Protection Against Allergic Disorders") discuss how helminthes infections result in the formation of IgE antibodies that recognize both the parasite and the allergen, but have low pathogenic potential. Interestingly, there is evidence of important differences in the allergen epitopes recognized by IgE from parasite-infected patients compared to allergic patients, and further research in this area holds the promise of exciting advances in our understanding of the mechanisms of allergy in the near future.

The novel ideas emerging from these studies suggest new preventive and therapeutic approaches based on interventions that alter IgE concentration and its affinity for allergens.

Alongside protection from parasites, IgE-mediated mast cell responses appear to be an important component of our bodies' anti-tumor defenses. Lai Sum Leoh and collaborators ("IgE Immunotherapy Against Cancer") explain the rationale behind the innovative use of IgE rather than IgG antibodies in cancer treatment, and discuss the therapeutic IgE antibodies that are currently under development for treatment of malignancies, as well as the possibility of prophylactic use of anti-tumor IgE and the challenges of effective delivery.

We hope this volume on IgE antibodies will help to disseminate current knowledge on the generation and function of IgE, and encourage further studies. It is also our aim to showcase the broad spectrum of IgE functions: in causing, as well as preventing, allergic diseases; in defense against parasites, and as a valuable component in protection from cancer. Only by taking all these facets of the complex biology of IgE into account can we hope to fully understand the workings of this intriguing biological molecule and so harness its potential to improve human health.

Juan J. Lafaille
Maria A. Curotto de Lafaille

References

Ishizaka K, Ishizaka T (1967) Identification of gamma-E-antibodies as a carrier of reaginic activity. J Immunol 99:1187–1198
Kinet JP (1999) The high-affinity IgE receptor (Fc epsilon RI): from physiology to pathology. Annu Rev Immunol 17:931–972
Muramatsu M et al. (2000) Class switch recombination and hypermutation require activation-induced cytidine deaminase (AID), a potential RNA editing enzyme. Cell 102:553–563

Contents

Biology of IgE Production: IgE Cell Differentiation and the Memory of IgE Responses... 1
Jin-Shu He, Sriram Narayanan, Sharrada Subramaniam, Wen Qi Ho, Juan J. Lafaille and Maria A. Curotto de Lafaille

Molecular Mechanisms of IgE Class Switch Recombination......... 21
Pei Tong and Duane R. Wesemann

Anti-IgE Therapy: Clinical Utility and Mechanistic Insights........ 39
Stephanie L. Logsdon and Hans C. Oettgen

New Insights on the Signaling and Function of the High-Affinity Receptor for IgE... 63
Ryo Suzuki, Jörg Scheffel and Juan Rivera

Helminth-Induced IgE and Protection Against Allergic Disorders.... 91
Firdaus Hamid, Abena S. Amoah, Ronald van Ree and Maria Yazdanbakhsh

IgE Immunotherapy Against Cancer........................... 109
Lai Sum Leoh, Tracy R. Daniels-Wells and Manuel L. Penichet

Index.. 151

Biology of IgE Production: IgE Cell Differentiation and the Memory of IgE Responses

Jin-Shu He, Sriram Narayanan, Sharrada Subramaniam, Wen Qi Ho, Juan J. Lafaille and Maria A. Curotto de Lafaille

Abstract The generation of long-lived plasma cells and memory B cells producing high-affinity antibodies depends on the maturation of B cell responses in germinal centers. These processes are essential for long-lasting antibody-mediated protection against infections. IgE antibodies are important for defense against parasites and toxins and can also mediate anti-tumor immunity. However, high-affinity IgE is also the main culprit responsible for the manifestations of allergic disease, including life-threatening anaphylaxis. Thus, generation of high-affinity IgE must be tightly regulated. Recent studies of IgE B cell biology have unveiled two mechanisms that limit high-affinity IgE memory responses: First, B cells that have recently switched to IgE production are programmed to rapidly differentiate into plasma cells, and second, IgE germinal center cells are transient and highly apoptotic. Opposing these processes, we now know that germinal center-derived IgG B cells can switch to IgE production, effectively becoming IgE-producing plasma cells. In this chapter, we will discuss the unique molecular and cellular pathways involved in the generation of IgE antibodies.

J.-S. He · S. Narayanan · S. Subramaniam · W.Q. Ho · M.A. Curotto de Lafaille (✉)
Singapore Immunology Network (SIgN), Agency for Science, Technology and Research, Singapore, Singapore
e-mail: maria_lafaille@immunol.a-star.edu.sg

S. Subramaniam
Nanyang Technological University, School of Biological Sciences, Singapore, Singapore

J.J. Lafaille
Skirball Institute and Department of Pathology, New York University School of Medicine, New York, USA

© Springer International Publishing Switzerland 2015
J.J. Lafaille and M.A. Curotto de Lafaille (eds.), *IgE Antibodies: Generation and Function*, Current Topics in Microbiology and Immunology 388,
DOI 10.1007/978-3-319-13725-4_1

Abbreviations

AID	Activation-induced cytidine deaminase
BCR	B cell receptor
CSR	Class switch recombination
EMPD	Extra membrane proximal domain
GC	Germinal center
LN	Lymph node
PC	Plasma cell
TLR	Toll-like receptor

Contents

1	Introduction	2
2	Regulation of Class Switch Recombination to IgE by Cytokines and Transcription Factors	3
3	T Cell-dependent Versus T Cell-independent Class Switching to IgE	4
4	Which T Cells Provide Help for B Cell Class Switching to IgE?	5
5	When and Where Does B Cell Class Switching to IgE Occur	6
6	Differential Regulation of the Production of IgE and IgG1	6
7	Direct Versus Sequential Switching to IgE and Their Biological Role	8
8	The Memory IgE Response	9
	8.1 Germinal Center IgE B Cells	9
	8.2 IgE Plasma Cells	10
	8.3 Mechanisms of IgE Memory	11
9	The Role of Membrane IgE in Cell Survival and Cell Fate	11
10	Conclusions	14
	References	15

1 Introduction

IgE antibodies are key mediators of allergic diseases, which manifest as acute allergic reactions such as hives, bronchospasm and systemic anaphylaxis, or chronic diseases such as rhinitis, atopic dermatitis, and asthma (Gould and Sutton 2008). The pathogenesis of IgE results from its ability to bind to high-affinity FcεRI receptors on the surface of mast cells and basophils (Kinet 1999). Crosslinking of surface-bound IgE by allergens causes FcεRI clustering, leading to mast cell degranulation and thereby the release of inflammatory mediators responsible for smooth muscle contraction, increased vascular permeability, secretion of mucus by epithelial cells, and inflammation (Galli and Tsai 2012).

The production of IgE is tightly regulated at the molecular and cellular level, resulting in a serum concentration of IgE that is several orders of magnitude lower than that of either IgG or IgA (Geha et al. 2003; Vieira and Rajewsky 1988). Recent studies have identified unique features of the differentiation of IgE-producing cells that account for its relative scarcity: Unlike the cells producing IgG, recently switched IgE B cells often rapidly differentiate into plasma cells (PCs) (Erazo et al. 2007; He et al. 2013; Yang et al. 2012); alongside, IgE-producing germinal center (GC) cells undergo high levels of apoptosis (He et al. 2013). However, the capacity for generation of high-affinity IgE can be maintained through a low level of switching of GC-derived IgG cells to IgE (Erazo et al. 2007; He et al. 2013; Xiong et al. 2012b). Here, we discuss the biological mechanisms involved in the generation of IgE responses, including the current knowledge of IgE GC cells and PCs, and the unique role of sequential switching in the IgE memory response.

2 Regulation of Class Switch Recombination to IgE by Cytokines and Transcription Factors

The generation of IgG, IgA, and IgE-producing cells occurs following activation of IgM- and IgD-expressing B cells via a process termed class switch recombination (CSR). When CSR to IgE occurs after initial recombination to an upstream isotype (usually IgG), the process is referred to as "sequential switching" (Mandler et al. 1993; Mills et al. 1992; Siebenkotten et al. 1992; Yoshida et al. 1990). CSR is induced in B cells after activation by ligation of surface receptors such as CD40, members of the toll-like receptor (TLR), or BAFF receptor family, in synergy with IL-4R receptor signaling (Geha et al. 2003). Activation induces expression of the enzyme activation-induced cytidine deaminase (AID) that mediates recombination between two DNA switch (S) regions, resulting in the joining of the targeted DNA sequences and deletion of intervening sequences (Muramatsu et al. 2000). S regions contain imperfect G-rich repeats that are located upstream of the Ig constant region (C) genes of IgM, all IgGs, IgA, and IgE. CSR is preceded by, and depends upon, transcription of the S regions, which is initiated in upstream intervening (I) germline promoters. Germline transcription results in the production of germline transcripts, in which a short I-exon is joined to the constant region, splicing out the switch region.

Activation of the germline I promoters is differentially regulated by transcription factors that are themselves activated by cytokine receptor signaling: CSR to IgE occurs downstream of IL-4 and IL-13 signaling and is inhibited by IFNγ, TGFβ, and IL2 (Coffman et al. 1993; de Vries et al. 1993; Finkelman et al. 1990; Punnonen et al. 1993; Suto et al. 2002). Signaling through the IL-4R triggers phosphorylation and nuclear translocation of STAT6. Binding of pSTAT6 to the Iε promoter is essential for promoter activation. pSTAT6 also induces the transcription factor NFIL3 (Rothman 2010), which binds to the Iε promoter and positively regulates transcription (Kashiwada et al. 2010).

Other transcription factors including E2A, NF-κB, PU.1, Pax5, AP-1, and C/EBP synergize with pSTAT6 to activate germline Iε transcription. In contrast, the transcriptional inhibitor BCL6, which is expressed in GC B cells, inhibits CSR to IgE (Harris et al. 1999). The transcription factor Id2, induced downstream of TGFβ signaling, is also a negative regulator of the Iε promoter (Sugai et al. 2003).

In sum, CSR to IgE depends on general mechanisms of B cell activation and signaling through the IL-4/IL-13 receptor. The fine regulation of CSR to IgE by cytokines secreted by CD4 T cells makes IgE responses highly sensitive to the T cell cytokine environment.

The molecular mechanisms involved in the generation of antibody genes and the specific features of immunoglobulin class switching to IgE are reviewed in detail by **Tong and Vesseman in this volume**.

3 T Cell-dependent Versus T Cell-independent Class Switching to IgE

The interaction between CD40 on the B cell surface and CD40L expressed on $CD4^+$ T cells is essential for antigen-specific IgE responses in vivo. The additional requirement for the T cell cytokines IL-4/IL-13 for B cell activation further highlights the dependence of IgE responses on T cell help.

Following engagement of B cell CD40 by CD40L, TRAF signaling factors are recruited and activated (Geha et al. 2003). TRAF activation ultimately results in nuclear translocation of the transcription factors NF-κB and AP-1. The engagement of CD40 also induces the activation of MAP kinase p38 (Faris et al. 1993, 1994; Pesu et al. 2002; Zhang et al. 2002). p38 is thought to directly regulate the activity of the transactivation domain of STAT6 (Pesu et al. 2002). The stimulation of IL-4R and CD40 thus has synergistic effects that enhance both germline Cε transcription and AID transcription through NF-κB and STAT6 (Geha et al. 2003).

Despite the obvious importance of T cell support for CSR to IgE, this requirement can be circumvented in the case of some B cell stimulants including TLR ligands such as lipopolysaccharide (LPS) (Snapper et al. 1988), or via BAFF/APRIL stimulation (Litinskiy et al. 2002), in the presence of IL-4. Sustained activation of NF-κB driven by TRIF is thought to be essential for the activation of the Cε locus and class switching to IgE driven by LPS plus IL-4 (Janssen et al. 2014).

In the case of BAFF and APRIL produced by dendritic cells (DCs), the outcome of B cell stimulation appears to depend on the local cytokine milieu; in the presence of IL-10, class switching to IgG is promoted, whereas IgE class switch is driven in the presence of IL-4 (Litinskiy et al. 2002).

Recently, the Yonehara group identified novel type 2 innate lymphoid cells (ILCs) in the spleen with the ability to enhance IgE production from murine B cells (Fukuoka et al. 2013). The mechanism underlying such enhancement is unclear, as is the physiological role of this process; since the ILCs are unable to produce IL-4, they are unlikely to be primary drivers of IgE class switching, though further study is clearly warranted. The Cerutti group found that ILCs in the spleen can integrate stromal and immunological signals to enhance antibody production by marginal zone B cells in both human and mouse (Magri et al. 2014). In this case, splenic ILCs support marginal zone B cells via provision of BAFF (or APRIL), CD40L, and NOTCH ligand Delta-like 1 (Magri et al. 2014). It would be worth assessing whether the novel type 2 splenic innate immunocyte described by the Yonehara group also produces BAFF (or APRIL) and CD40L and to further elucidate their possible involvement in the enhancement of IgE production.

The relevance of CD40-independent pathways of IgE production in vivo is not known. They could be involved in the production of IgE associated with some immunodeficiencies (Xiong et al. 2012a).

4 Which T Cells Provide Help for B Cell Class Switching to IgE?

Early studies of IgE responses in immunized mice or in mice infected with helminth parasites demonstrated that, in these systems, the production of IgE was dependent on $CD4^+$ T cells and the cytokine IL-4 (Finkelman et al. 1988; Katona et al. 1988). Furthermore, IL-4 was shown to be necessary for the production of IgE, but not IgG1, in secondary responses (Katona et al. 1991), implying an absence of memory IgE cells and therefore that de novo class switching to IgE occurs in recall, as well as primary responses.

The generation of mice expressing reporter genes in the IL-4 locus (Mohrs et al. 2005) has facilitated the characterization of IL-4-producing cells in vivo. Two populations of IL-4-producing cells in lymphoid organs, Tfh and Th2-like cells, in addition to tissue Th2 cells, are potential helper cells for B cell class switching to IgE. Isolation of T–B cell conjugates from LNs of IL-4 reporter mice showed higher levels of post-switched IgG1 transcripts in conjugates of B cells with IL-4 producing Tfh cells than in conjugates of B cells with IL-4 producing non-Tfh cells. Upon adoptive transfer, both populations of IL-4 producing T helper cells were able to mediate class switching to IgG1. The production of IgE was not analyzed (Reinhardt et al. 2009). Although the results indicate that both Tfh and Th2-like cells from LN can mediate CSR to IgG1, they cannot be taken as evidence that either population is responsible for CSR to IgE.

Recent studies by Turqueti Neves et al. showed that in bone marrow chimeric mice in which IL-4 and IL-13 could only be produced by T helper cells unable to localize to the B cell follicles (due to CXCR5 deficiency), neither GC formation, nor the production of IgG1 and IgE were compromised. This suggested that IL-4/IL-13 produced by non-Tfh cells is sufficient for CSR to IgE and IgG1 (Turqueti-Neves et al. 2014). In contrast to these data, mice with a deletion in the CNS2 DNA control element of the IL-4 locus, which is necessary for IL-4 expression in Tfh but not for expression in in-vitro differentiated Th2 cells, were highly deficient in the production of both IgG1 and IgE. This apparent contradiction might be explained if the IL-4-producing non-Tfh cells of the spleen and LNs require an active CNS2 locus for IL-4 expression, unlike in vitro-differentiated Th2 cells. Furthermore, it was shown that Th2 cells could differentiate into Tfh cells in vivo (Glatman Zaretsky et al. 2009). Thus, the data so far are not conclusive as to which T helper population mediates CSR to IgE.

5 When and Where Does B Cell Class Switching to IgE Occur

Related to the issue of T cell help for IgE production, the timing and tissue environment in which CSR to IgE takes place is also not yet well understood. Kinetically, the production of IgE in T-dependent responses is associated with the germinal center phase and not with the initial extra-follicular antibody production (Erazo et al. 2007). The fast decline of IgE GC cells and IgE PC in primary responses (He et al. 2013; Yang et al. 2012) suggests that most CSR to IgE occurs in early GC phase and not in the mature GC. CSR to IgE may thus take place as activated B cells interact with Th2-like or pre-Tfh cells during the early GC reaction. The expression of BCL6 in GC B cells, as well as the production of IL-21 by Tfh cells are potential inhibitor of CSR to IgE in the GC (Harris et al. 2005; Ozaki et al. 2002; Suto et al. 2002; Vinuesa et al. 2005). Nevertheless, heterogeneity in BCL6 expression by GC B cells and heterogeneity in IL-21 and IL-4 production by Tfh cells may create a permissive environment for class switching to IgE in the GC.

There is ample evidence that class switching to IgE occurs in the nasal mucosa of allergic rhinitis patients (Cameron et al. 2003; Coker et al. 2003; Takhar et al. 2005), patients with chronic rhinosinusitis with nasal polyps (Gevaert et al. 2013), and in the bronchial mucosa of asthmatic individuals (Takhar et al. 2007). Detection of Sμ-Sε and Sγ-Sε switch DNA circles in the nasal mucosa (Cameron et al. 2003), and Iε-Cμ and Iε-Cγ circle transcripts in bronchial mucosa (Takhar et al. 2007) indicates the occurrence of direct and sequential switching. It is not known if class switching to IgE in the respiratory mucosa is necessarily associated with organized lymphoid tissues, but GC-like structures were detected in the nasal mucosa of rhinitis patients and in nasal polyps (Gevaert et al. 2013), as well as in the lungs of mice with chronic allergic inflammation (Curotto de Lafaille et al. 2008). Allergen-specific IgE was found to constitute a much higher proportion of total IgE in the nasal mucosa than in the blood of patients with grass pollen allergy, implying that local IgE production is important for mucosal mast cell sensitization and localized allergic reactions (Smurthwaite et al. 2001).

6 Differential Regulation of the Production of IgE and IgG1

Stimulation of mouse B lymphocytes through TLRs or CD40 in the presence of IL-4 promotes CSR to IgE and IgG1, and in reality, far more B cells switch to production of IgG1 under these conditions than to IgE, both in vivo and in vitro

(Rothman et al. 1988; Siebenkotten et al. 1992). This may reflect the fact that the ε switch region of IgE is much shorter and less repetitive than the γ1 switch region, which decreases the statistical probability of the recombinase AID binding to this section of DNA (Hackney et al. 2009). In addition, while the Iγ1 germline promoter is activated by the transcription factor ATF-2/CREB2, which is upregulated in a sustained fashion during B cell stimulation, Iε transcripts rely upon AP-1 for their activation, which is upregulated only transiently (Mao and Stavnezer 2001).

SWAP-70 is another molecule involved in differential regulation of IgE and IgG class switching. SWAP-70-deficient mice have impaired IgE production, while LPS-induced B cell proliferation and switching to other isotypes, including IgG1, is normal (Borggrefe et al. 2001). Recently, it was shown that SWAP-70 binds to Iε promoter and is required for association of STAT6 with the Iε promoter, and that in the absence of SWAP-70, the occupancy of inhibitory BCL6 on the Iε promoter is substantially increased (Audzevich et al. 2013). Thus, SWAP-70 mediates IgE production through regulation of antagonistic STAT6 and BCL6 occupancy of the Iε promoter.

Cytokines also have important roles in the differential regulation of CSR to IgE and IgG1. Following IL-4/IL-13 signaling and STAT6 activation, IgE CSR is favored by the germline ε promoter region, which binds STAT6 with a 10-fold higher affinity than the γ1 germline promoter (Mao and Stavnezer 2001). Therefore, Cε germline transcription is far more responsive to IL-4 than the γ1 promoter, and accordingly, IgG1 CSR can take place in the absence of IL-4. In contrast, IL-21, which is produced by T cells in GCs, promotes the production of IgG1 and at the same time inhibits CSR to IgE (Erazo et al. 2007; Ozaki et al. 2002). Other cytokines, such as TGF-β and IFN-γ, act directly on B cells to repress both Cε and Cγ1 germline transcription (Geha et al. 2003).

Adding a further level of complexity to the regulation of IgE production is the activity of various kinases/phosphatases. Phosphoinositide 3-kinase (PI3K) negatively regulates IgE production at both CSR and protein levels in vivo and in vitro (Doi et al. 2008) by maintaining levels of BCL6 that preclude STAT6 binding to the Iε promoter (Zhang et al. 2008, 2012). The transmembrane phosphatase CD45 can inhibit IL-4-mediated CSR to IgE in human B cells (Loh et al. 1995) by blocking activation of STAT6 through its JAK phosphatase activity or by inhibiting CD40-initiated activation of c-Jun NH2-terminal kinase and p38 (Arimura et al. 2001) that are involved in the induction of ε germline transcription.

Taken together, it is evident that the process of CSR is tightly regulated by the complex interplay of several factors. The decision to switch to IgE production will therefore be governed by the timing, strength, and combination of signals received by the individual B cell, in a process that we are yet to fully understand.

7 Direct Versus Sequential Switching to IgE and Their Biological Role

Class switching to IgE may occur through direct recombination between the switch (S) regions of Cμ and Cε genes (Sμ and Sε, respectively) or by the sequential switching of B cells that have already undergone recombination to IgG (**Tong and Vesemann, this volume**). In the mouse, sequential switching occurs predominantly in cells with a Sμ-Sγ1 rearrangement (Siebenkotten et al. 1992). In humans, remnants of all the four Sγ regions (Sγ1, Sγ2, Sγ3, and Sγ4) have been found in hybrid Sμ-Sγ-Sε regions of IgE cells (Mills et al. 1992, 1995). While the presence of Sγ repeat remnants in the Sμ-Sε region of IgE cells indicates a sequential switching event, Sγ remnants may be lost in the secondary recombination, which means that the frequency of sequential switching is most likely underestimated.

Initial data from mice suggested that sequential switching was not required for IgE production since animals bearing a mutation in DNA regulatory sequences upstream of Sγ1, which suppressed IgG1 class switching, remained able to produce IgE (Jung et al. 1994). However, more detailed analysis in IgG1-deficient mice revealed that, in the absence of sequential switching, only low-affinity IgE was produced (Xiong et al. 2012b). From these studies, we now know that sequential switching is primarily important for the production of high-affinity IgE and is therefore predicted to play an important role in the pathogenesis of IgE-mediated allergic responses such as anaphylaxis (Xiong et al. 2012a).

In addition to the affinity of IgE for an allergen, another important factor for the pathogenic potential of IgE is the relative concentration of allergen-specific IgE in the total IgE pool. The affinity and concentration of IgE determine the probability that two or more allergen-specific IgE molecules will be cross-linked by a multivalent allergen on the mast cell surface. We showed that an excess of low-affinity IgE inhibited the passive cutaneous anaphylaxis reaction mediated by high-affinity IgE (Xiong et al. 2012b).

Accordingly, high levels of IgE in some conditions such as helminth infections are not associated with occurrence of allergic diseases even though patients harbor allergen-specific IgE antibodies. Interestingly, helminth-infected patients recognize cross-reactive determinants in peanut allergens that are different than the determinants recognized by peanut-allergic patients (**Hamid et al. this volume**). The high risk of adverse reaction to peanuts in allergic patients may thus be related to the specificity and affinity of the IgE antibodies, as well as the relatively high concentration of peanut-specific IgE in the total IgE pool.

In addition to their distinct biological roles in the generation of high- and low-affinity IgE, direct and sequential switching drives different B cell differentiation pathways (He et al. 2013). We found that the switch regions of GC IgE-producing cells were devoid of Sγ1 footprints, indicating that they originated from IgM-producing cells by direct switching. In contrast, the switch regions of IgE PCs consistently contained Sγ1 repeats, indicating that most or a large proportion of the IgE PC originated from sequential switching (He et al. 2013).

In sum, direct class switching to IgE is involved in the generation of GC IgE cells and linked to the production of low-affinity IgE, while sequential switching generates IgE PCs and is essential for the production of high-affinity IgE.

8 The Memory IgE Response

8.1 Germinal Center IgE B Cells

Primary antibody responses usually comprise an initial extra-follicular differentiation phase that generates short-lived PCs and a delayed response that involves the formation of GCs. It is in GCs that B lymphocytes proliferate, mutate their antigen receptors, and are positively selected through their high-affinity interactions with follicular DCs and Tfh cells (Tarlinton 2008; Victora and Nussenzweig 2012). GCs are essential for the generation of high-affinity antibody secreting long-lived PCs and memory B cells. Mature GCs have two distinct regions: the dark zone, in which cell division and somatic hyper-mutation take place, and a light zone, where positive and negative selection occurs. Positively selected light zone B cells may reenter the dark zone for another cycle of proliferation/mutation, to differentiate into PCs or to become memory cells.

Initial studies of IgE B cells in mice revealed that while IgE-producing cells were predominantly generated during the GC phase of the immune response, GC IgE B cells were not readily detectable in the spleen or LNs of these animals (Erazo et al. 2007). Detailed analysis showed that the IgE B cell population in these mice was dominated by PCs. Furthermore, sequence analysis of IgE antibodies after repeated immunization provided evidence of affinity maturation. It was thus proposed that class switching to IgE was rapidly followed by differentiation into PC and that in the absence of a clear IgE GC phase, high-affinity IgE was generated by sequential switching of GC-derived IgG1-expressing cells (Erazo et al. 2007).

However, the development of reporter mice for IgE expression revealed that bona fide GC IgE-producing cells do exist (He et al. 2013; Talay et al. 2012; Yang et al. 2012). Kinetic analysis of the frequency of GC IgE B cells revealed that they decreased in numbers very quickly, unlike IgG1 B cells, which persisted in the mature GC (He et al. 2013; Yang et al. 2012). One group attributed the decline of IgE GC cells to their differentiation into PCs (Yang et al. 2012, 2014). Our results however indicated that the intrinsic properties of IgE GC cells resulted in increased apoptosis and rapid decline of the population (He et al. 2013).

We found that IgE-producing GC cells expressed threefold to fourfold lower levels of surface immunoglobulin than IgG1 GC cells, had lower levels of proximal BCR signaling than IgG1 cells, and were unable to populate the light zone of the GC (He et al. 2013). Furthermore, most apoptotic IgE B cells had a dark zone phenotype, while IgG1 GC cells undergoing apoptosis exhibited mostly a light

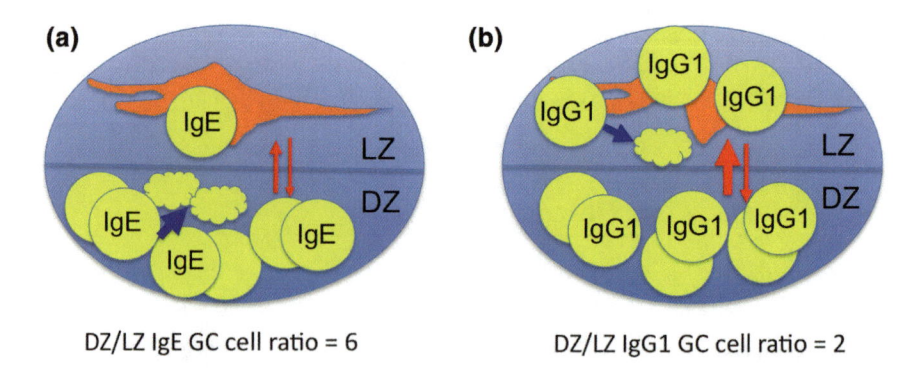

Fig. 1 Atypical distribution of IgE-producing B cells in the dark zone (*DZ*) and light zone (*LZ*) of the GC. **a** IgE GC cells are highly apoptotic and are underrepresented in the LZ of the GC. Most apoptotic IgE B cells in the GC have a DZ phenotype. **b** IgG1 GC cells are twice as frequent in the DZ than in the LZ. Most apoptotic IgG1 GC cells are in the LZ. The best-fit hypothesis generated by mathematical modeling predicts that the IgE DZ cells fail to migrate to the LZ and die in the DZ (He et al. 2013). *Blue arrows* indicate the apoptotic pathway. *Red arrows* represent the intensity of the inter-zonal migration pattern

zone phenotype (Fig. 1). We used mathematical modeling of the GC reaction (Meyer-Hermann et al. 2012) to explain the phenotype of IgE GC cells; the best-fit hypothesis predicted that the GC dark zone IgE cells were somehow unable to exit the dark zone and become light zone cells and thus most would die in the dark zone (Fig. 1). Since IgE B cells in the GC have low BCR expression and signaling, we believe that they may fail at a BCR-dependent checkpoint after expression of a new mutated receptor or in the transition from the dark zone to the light zone (He et al. 2013).

8.2 IgE Plasma Cells

In mice, IgE-producing PCs differ from those of other isotypes in that they express higher levels of surface immunoglobulin (He et al. 2013; Yang et al. 2012). In addition, IgE-expressing PCs constitute 20–40 % of all spleen and LN PCs during Th2 responses despite the fact that antigen-specific serum IgE levels are many fold lower than IgG1 levels (Erazo et al. 2007; He et al. 2013; Yang et al. 2012). There is evidence that IgE-producing PCs are less responsive to stromal cell-derived factor 1 than IgG-expressing PCs, which may impair their ability to localize to the bone marrow (Achatz-Straussberger et al. 2008). However, IgE-producing PCs were detected in the bone marrow after immunization or helminth infection (He et al. 2013; Luger et al. 2009).

8.3 Mechanisms of IgE Memory

B cell memory has been studied predominantly in the case of IgG responses, which are typically maintained by classical memory B cells that respond rapidly to antigen-mediated activation and by long-lived PCs that reside in the spleen and bone marrow. Until recently, the B cell population responsible for IgE production in secondary responses was unknown.

We have now demonstrated that B220$^+$CD138$^-$IgG1$^+$ B cells, which contain both GC and memory IgG1-expressing cell populations, accounted for the bulk of IgE production in recall responses. In contrast, B220$^+$CD138$^-$IgE$^+$ B cells made little or no contribution to the secondary IgE response (He et al. 2013). These results provide an explanation for the established observation of the need for de novo class switching in IgE memory responses (Finkelman et al. 1988). The data are also supported by earlier findings by Nojima and collaborators, where in vitro-generated GC phenotype (iGB) IgG1 and IgE cells were adoptively transferred into mice, revealing that IgG1 iGB cells, but not IgE iGB cells, gave rise to memory B cells capable of mounting a T cell-dependent immune response (Nojima et al. 2011).

In contrast to the observations described above, L. Wu's group using M1′GFP mice reported that IgE GC cells, but not IgG1 cells, generate IgE memory cells and IgE PCs (Talay et al. 2012). Further analysis by the same group concluded that IgG1 memory cells, in addition to IgE memory cells, could generate IgE antibodies (Talay et al. 2013; Wu and Zarrin 2014). However, concerns remain that modifications made to the IgE gene in the M1′GFP mouse may have altered the level of production of membrane IgE and the biological proprieties of IgE-producing cells (Lafaille et al. 2012; Xiong et al. 2012a).

Another important aspect of B cell memory is the long-term production of antibodies by PCs. We studied the ability of IgE- and IgG1-secreting PCs to mediate antibody production following adoptive transfer and found that CD138$^+$IgE$^+$ PCs but not CD138$^+$IgG1$^+$ PCs from lymphoid organs transferred the production of IgE antibodies to recipient mice (He et al. 2013). This showed that IgE PCs are important to sustain the memory of IgE responses and that IgG1 PCs are too differentiated to undergo sequential switching to IgE.

In summary, IgE memory responses in mice are maintained by a combination of the sequential switching of IgG1-producing B cells to IgE production and by IgE PCs. A model illustrating the production of IgE antibodies in mice is shown in Fig. 2.

9 The Role of Membrane IgE in Cell Survival and Cell Fate

Immunoglobulins are produced either in a membrane-bound form or as soluble secreted proteins through a process of differential polyadenylation and splicing (Edwalds-Gilbert et al. 1997; Venkitaraman et al. 1991). The membrane form is the main form expressed in B lymphocytes. On the other hand, for most isotypes, the differentiation of GC or memory B cells into PCs is accompanied by a switch

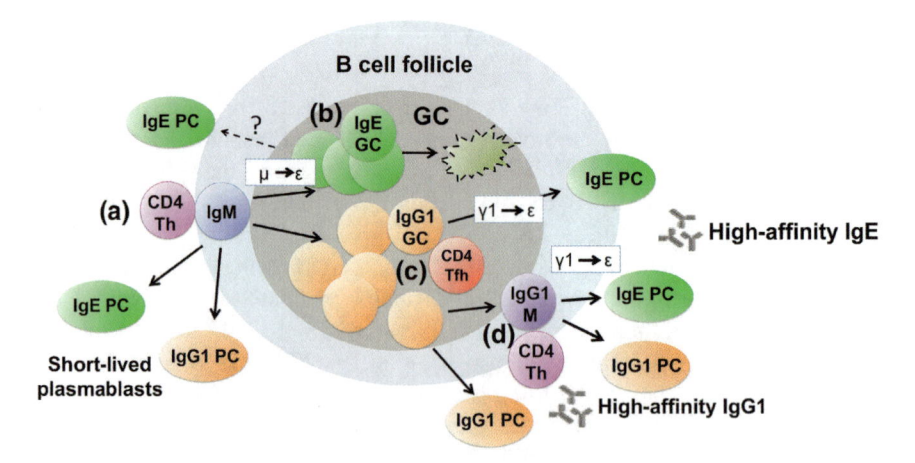

Fig. 2 Model for the generation of low- and high-affinity IgE antibodies. **a** Encounters between activated IgM B cells and IL-4-producing CD4 T helper cells (*CD4 Th*) in the T-B zone border generate short-lived IgG1 and IgE PC, and IgG1 and IgE GC cells. IgE-producing GC cells are generated through direct switching (μ → ε). **b** IgE GC cells are highly apoptotic and do not give rise to high-affinity IgE PC or IgE memory cells. Some early IgE GC cells may generate short-lived IgE-secreting PC (*dashed arrows*). **c** IgG1 GC cells differentiate into high-affinity IgG1 PC and IgG1 memory cells. IgG1 GC cells interacting with IL-4-producing CD4 Tfh cells may undergo sequential switching (γ1 → ε) and generate high-affinity IgE PC. **d** IgG1 memory cells (IgG1 M) differentiate into IgG1 PC or sequentially switch to IgE production and give rise to IgE PC after interaction with IL-4-producing CD4 Th cells

to the predominant production of soluble immunoglobulin. However, the expression of IgE differs from the expression of other membrane immunoglobulin isotypes in multiple ways (Laffleur et al. 2014). Membrane IgE is expressed at low levels in B220$^+$IgE B cells and is upregulated in CD138$^+$IgE PCs (He et al. 2013; Yang et al. 2012). In addition, human and mouse B220$^+$IgE B cells express relatively high levels of the soluble IgE (He et al. 2013; Karnowski et al. 2006). The low membrane/secreted IgE ratio results from inefficient polyadenylation of membrane IgE. In both human and mouse IgE genes, the internal polyadenylation signal for secreted IgE has the consensus AATAAA sequence, while three deviant polyadenylation signals (AGTAA, AAGAAA, and ATTAAA) are present in the 3′ untranslated region downstream of the membrane exons (Fig. 3). These atypical polyadenylation signals dramatically decrease the efficiency of polyadenylation for the IgE membrane transcript (Karnowski et al. 2006) and thereby account for the preferential production of soluble IgE by B220$^+$B cells.

The membrane and cytoplasmic regions of IgE are encoded by two membrane exons, M1 and M2. Human M1 exon encodes for a unique long extra membrane proximal domain (EMPD) not found in mice. Human IgE can also carry a short EMPD produced by alternative splicing (Fig. 3). Ectopic expression of human IgE containing the short or long EMPD domains in mouse B cells suggested a proliferation-inhibition and pro-apoptotic function of IgE containing the short EMPD (Batista et al. 1996; Poggianella et al. 2006).

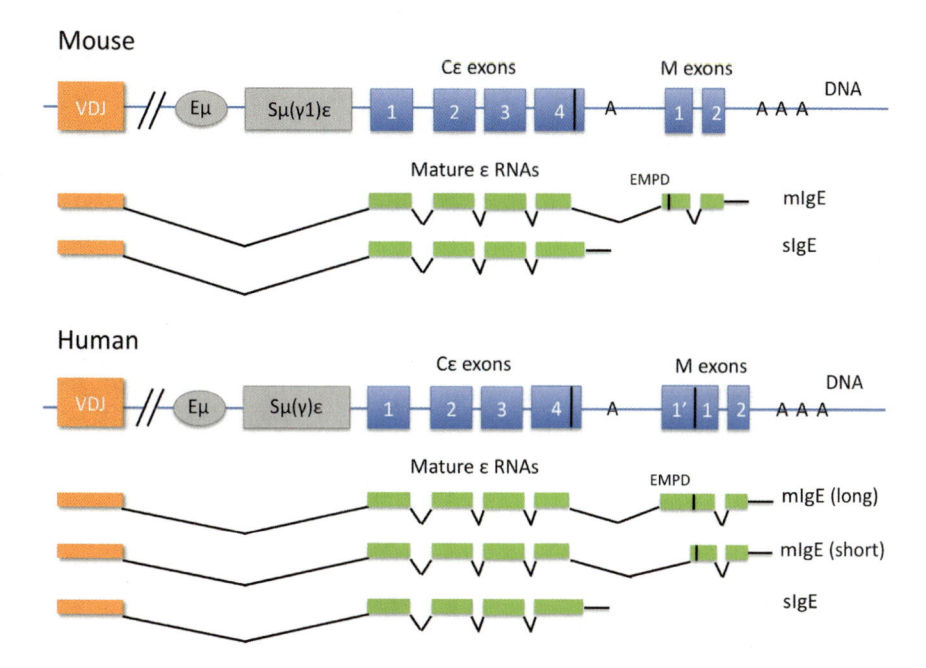

Fig. 3 The structure of mouse and human mature IgE transcripts. Graphic representation of the structure of the mouse and human IgE heavy chain genes (DNA) and mature transcripts. The Cε region comprises four exons, each encoding an immunoglobulin-like domain. The two membrane (*M*) exons encode the extra membrane proximal domain (*EMPD*), the transmembrane region, and a cytoplasmic tail. In humans, the M1 exon carries additional 5′ DNA sequences (M1′) that encode for 52 extra amino acids in the EMPD. Human mIgE can be produced with a short or long EMPD, while mouse IgE has only a short EMPD. The membrane and secreted forms of IgE (mIgE and sIgE, respectively) are generated through the use of different polyadenylation sites (A) and alternative splicing. The internal polyadenylation site for secreted IgE has a consensus signal, while three atypical polyadenylation sites are located downstream of the M exons. Inefficiency in 3′ polyadenylation leads to preferential production of sIgE in IgE B cells. Em: enhancer of m; Sμ(γ1)ε and Sμ(γ)ε: switch (S) region of the IgE genes containing or lacking Sγ repeats

The mouse IgE cytoplasmic tail was found to bind the anti-apoptotic molecule HAX-1 (Oberndorfer et al. 2006). In transfected cells expressing full-length IgE heavy chain, but not IgE lacking the cytoplasmic tail, HAX-1 promoted the internalization of membrane IgE. This would explain the lower IgE serum levels in mice expressing IgE lacking the cytoplasmic tail (Achatz et al. 1997). It was also speculated that association with HAX-1 is required for antigen presentation by IgE B cells to T cells (Oberndorfer et al. 2006).

Recent studies have also made interesting observations on the peculiarities of the structure of IgE antibodies. The constant region of IgE consists of four immunoglobulin-type domains, Cε1 to Cε4 (Fig. 3), which are similar to IgM, but different from the IgG constant region that is formed by three Cγ domains and a hinge region. Furthermore, in IgE, the disulfide-linked Cε2 domain replaces the hinge, and analysis of the crystal structure of the soluble Cε2-Cε4 dimer (IgE-Fc) demonstrated that the IgE molecule is asymmetrically bent, with the Cε2 domains folded back into the Cε3 and Cε4 domains (Wan et al. 2002). An acutely bent conformation was also detected for IgE bound to FcεRIα (Holdom et al. 2011). This conformation of the IgE-Fc region appears to be responsible for the slow dissociation rate of IgE from FcεRI, as it facilitates the interaction of both Cε3 domains with the two binding sites in FcεRIα (Holdom et al. 2011). It was proposed that membrane-bound IgE also has a bent conformation that causes the Fab fragments to face toward the membrane, in a non-favorable position for antigen binding. Recently, however, it was shown that IgE can flip through an extended conformation that may enable antigen recognition (Drinkwater et al. 2014). Whether membrane-bound IgE adopts a mostly bent or flexible structure could have profound impacts on its cellular function in terms of antigen binding and BCR signaling.

In summary, IgE genes, mRNA expression, and protein structure have marked differences from that of other immunoglobulins. These characteristics are likely to be involved in the unique aspects of IgE cell function, survival, and fate.

10 Conclusions

The development of new models to study IgE in mice, alongside methodological advances in cellular and molecular analysis, human genetics, and bioinformatics, have paved the way for an unprecedented depth of study of IgE responses that has already yielded some exciting insights into the biology of IgE. Furthermore, interest in IgE has been renewed by the success of anti-IgE therapies, which have cemented the role of IgE in chronic inflammatory allergic diseases. As such, further studies of IgE are necessary and hold the promise of clinical advancement of therapeutics.

The production of IgE antibodies is tightly controlled in mice and humans through several mechanisms that include suppression of differentiation of Th2 cells by regulatory T cells and cytokines, characteristics of the switch region of IgE that lead to low recombination frequency, increased apoptosis of GC IgE cells, impaired formation of IgE memory cells, and high turnover of IgE antibodies. Nevertheless, high-affinity IgE antibodies can form, albeit in limited amounts, through the sequential switching of GC-derived IgG cells with the necessary help of T cells and IL-4. The high threshold required for production of high-affinity IgE suggests that strong evolutionary pressure operates to limit, but not eliminate, the production of IgE. This serves to reinforce the notion that the biological activity of IgE must be constrained to avoid immunopathology, while retaining the ability to exploit its protective capacities during infection.

Acknowledgments MACL laboratory is supported by A*STAR-SIgN core funding and by the Joint Council Organization (JCO) grant 1431AFG104, Singapore. J.J. Lafaille laboratory is supported by the Multiple Sclerosis Society, a B Levine scholarship and the NIH. We wish to thank Dr Lucy Robinson of Insight Editing London for her assistance in the preparation of the text.

References

Achatz G, Nitschke L, Lamers MC (1997) Effect of transmembrane and cytoplasmic domains of IgE on the IgE response. Science 276:409–411

Achatz-Straussberger G, Zaborsky N, Konigsberger S, Luger EO, Lamers M, Crameri R, Achatz G (2008) Migration of antibody secreting cells towards CXCL12 depends on the isotype that forms the BCR. Eur J Immunol 38:3167–3177

Arimura Y, Ogimoto M, Mitomo K, Katagiri T, Yamamoto K, Volarevic S, Mizuno K, Yakura H (2001) CD45 is required for CD40-induced inhibition of DNA synthesis and regulation of c-Jun NH2-terminal kinase and p38 in BAL-17 B cells. J Biol Chem 276:8550–8556

Audzevich T, Pearce G, Breucha M, Gunal G, Jessberger R (2013) Control of the STAT6-BCL6 antagonism by SWAP-70 determines IgE production. J Immunol 190:4946–4955

Batista FD, Anand S, Presani G, Efremov DG, Burrone OR (1996) The two membrane isoforms of human IgE assemble into functionally distinct B cell antigen receptors. J Exp Med 184:2197–2205

Borggrefe T, Keshavarzi S, Gross B, Wabl M, Jessberger R (2001) Impaired IgE response in SWAP-70-deficient mice. Eur J Immunol 31:2467–2475

Cameron L, Gounni AS, Frenkiel S, Lavigne F, Vercelli D, Hamid Q (2003) Sε Sμ and Sε Sγ switch circles in human nasal mucosa following ex vivo allergen challenge: evidence for direct as well as sequential class switch recombination. J Immunol 171:3816–3822

Coffman RL, Lebman DA, Rothman P (1993) Mechanism and regulation of immunoglobulin isotype switching. Adv Immunol 54:229–270

Coker HA, Durham SR, Gould HJ (2003) Local somatic hypermutation and class switch recombination in the nasal mucosa of allergic rhinitis patients. J Immunol 171:5602–5610

Curotto de Lafaille MA, Kutchukhidze N, Shen S, Ding Y, Yee H, Lafaille JJ (2008) Adaptive Foxp3 + regulatory T cell-dependent and -independent control of allergic inflammation. Immunity 29:114–126

de Vries JE, Punnonen J, Cocks BG, de Waal Malefyt R, Aversa G (1993) Regulation of the human IgE response by IL4 and IL13. Res Immunol 144:597–601

Doi T, Obayashi K, Kadowaki T, Fujii H, Koyasu S (2008) PI3K is a negative regulator of IgE production. Int Immunol 20:499–508

Drinkwater N, Cossins BP, Keeble AH, Wright M, Cain K, Hailu H, Oxbrow A, Delgado J, Shuttleworth LK, Kao MW et al (2014) Human immunoglobulin E flexes between acutely bent and extended conformations. Nat Struct Mol Biol 21:397–404

Edwalds-Gilbert G, Veraldi KL, Milcarek C (1997) Alternative poly(A) site selection in complex transcription units: means to an end? Nucleic Acids Res 25:2547–2561

Erazo A, Kutchukhidze N, Leung M, Christ AP, Urban JF Jr, Curotto de Lafaille MA, Lafaille JJ (2007) Unique maturation program of the IgE response in vivo. Immunity 26:191–203

Faris M, Gaskin F, Geha RS, Fu SM (1993) Tyrosine phosphorylation defines a unique transduction pathway in human B cells mediated via CD40. Trans Assoc Am Phys 106:187–195

Faris M, Gaskin F, Parsons JT, Fu SM (1994) CD40 signaling pathway: anti-CD40 monoclonal antibody induces rapid dephosphorylation and phosphorylation of tyrosine-phosphorylated proteins including protein tyrosine kinase Lyn, Fyn, and Syk and the appearance of a 28-kD tyrosine phosphorylated protein. J Exp Med 179:1923–1931

Finkelman FD, Katona IM, Urban JF Jr, Holmes J, Ohara J, Tung AS, Sample JV, Paul WE (1988) IL-4 is required to generate and sustain in vivo IgE responses. J immunol 141:2335–2341

Finkelman FD, Holmes J, Katona IM, Urban JF Jr, Beckmann MP, Park LS, Schooley KA, Coffman RL, Mosmann TR, Paul WE (1990) Lymphokine control of in vivo immunoglobulin isotype selection. Annu Rev Immunol 8:303–333

Fukuoka A, Futatsugi-Yumikura S, Takahashi S, Kazama H, Iyoda T, Yoshimoto T, Inaba K, Nakanishi K, Yonehara S (2013) Identification of a novel type 2 innate immunocyte with the ability to enhance IgE production. Int Immunol 25:373–382

Galli SJ, Tsai M (2012) IgE and mast cells in allergic disease. Nat Med 18:693–704
Geha RS, Jabara HH, Brodeur SR (2003) The regulation of immunoglobulin E class-switch recombination. Nat Rev Immunol 3:721–732
Gevaert P, Nouri-Aria KT, Wu H, Harper CE, Takhar P, Fear DJ, Acke F, De Ruyck N, Banfield G, Kariyawasam HH et al (2013) Local receptor revision and class switching to IgE in chronic rhinosinusitis with nasal polyps. Allergy 68:55–63
Glatman Zaretsky A, Taylor JJ, King IL, Marshall FA, Mohrs M, Pearce EJ (2009) T follicular helper cells differentiate from Th2 cells in response to helminth antigens. J Exp Med 206:991–999
Gould HJ, Sutton BJ (2008) IgE in allergy and asthma today. Nat Rev Immunol 8:205–217
Hackney JA, Misaghi S, Senger K, Garris C, Sun Y, Lorenzo MN, Zarrin AA (2009) DNA targets of AID evolutionary link between antibody somatic hypermutation and class switch recombination. Adv Immunol 101:163–189
Harris MB, Chang CC, Berton MT, Danial NN, Zhang J, Kuehner D, Ye BH, Kvatyuk M, Pandolfi PP, Cattoretti G et al (1999) Transcriptional repression of Stat6-dependent interleukin-4-induced genes by BCL-6: specific regulation of iε transcription and immunoglobulin E switching. Mol Cell Biol 19:7264–7275
Harris MB, Mostecki J, Rothman PB (2005) Repression of an interleukin-4-responsive promoter requires cooperative BCL-6 function. J Biol Chem 280:13114–13121
He JS, Meyer-Hermann M, Xiangying D, Zuan LY, Jones LA, Ramakrishna L, de Vries VC, Dolpady J, Aina H, Joseph S et al (2013) The distinctive germinal center phase of IgE$^+$ B lymphocytes limits their contribution to the classical memory response. J Exp Med 210:2755–2771
Holdom MD, Davies AM, Nettleship JE, Bagby SC, Dhaliwal B, Girardi E, Hunt J, Gould HJ, Beavil AJ, McDonnell JM et al (2011) Conformational changes in IgE contribute to its uniquely slow dissociation rate from receptor FcεRI. Nat Struct Mol Biol 18:571–576
Janssen E, Ozcan E, Liadaki K, Jabara HH, Manis J, Ullas S, Akira S, Fitzgerald KA, Golenbock DT, Geha RS (2014) TRIF signaling is essential for TLR4-driven IgE class switching. J immunol 192:2651–2658
Jung S, Siebenkotten G, Radbruch A (1994) Frequency of immunoglobulin E class switching is autonomously determined and independent of prior switching to other classes. J Exp Med 179:2023–2026
Karnowski A, Achatz-Straussberger G, Klockenbusch C, Achatz G, Lamers MC (2006) Inefficient processing of mRNA for the membrane form of IgE is a genetic mechanism to limit recruitment of IgE-secreting cells. Eur J Immunol 36:1917–1925
Kashiwada M, Levy DM, McKeag L, Murray K, Schroder AJ, Canfield SM, Traver G, Rothman PB (2010) IL-4-induced transcription factor NFIL3/E4BP4 controls IgE class switching. Proc Natl Acad Sci USA 107:821–826
Katona IM, Urban JF Jr, Finkelman FD (1988) The role of L3T4 + and Lyt-2 + T cells in the IgE response and immunity to *Nippostrongylus brasiliensis*. J immunol 140:3206–3211
Katona IM, Urban JF Jr, Kang SS, Paul WE, Finkelman FD (1991) IL-4 requirements for the generation of secondary in vivo IgE responses. J Immunol 146:4215–4221
Kinet JP (1999) The high-affinity IgE receptor (FcεRI): from physiology to pathology. Annu Rev Immunol 17:931–972
Lafaille JJ, Xiong H, Curotto de Lafaille MA (2012) On the differentiation of mouse IgE+ cells. Nat Immunol 13:623 author reply 623–624
Laffleur B, Denis-Lagache N, Peron S, Sirac C, Moreau J, Cogne M (2014) AID-induced remodeling of immunoglobulin genes and B cell fate. Oncotarget 5:1118–1131
Litinskiy MB, Nardelli B, Hilbert DM, He B, Schaffer A, Casali P, Cerutti A (2002) DCs induce CD40-independent immunoglobulin class switching through BLyS and APRIL. Nat Immunol 3:822–829
Loh RK, Jabara HH, Ren CL, Fu SM, Vercelli D, Geha RS (1995) Role of protein tyrosine kinases and phosphatases in isotype switching: crosslinking CD45 to CD40 inhibits IgE isotype switching in human B cells. Immunol Lett 45:99–106

Luger EO, Fokuhl V, Wegmann M, Abram M, Tillack K, Achatz G, Manz RA, Worm M, Radbruch A, Renz H (2009) Induction of long-lived allergen-specific plasma cells by mucosal allergen challenge. J Allergy Clin Immunol 124:819–826 e814

Magri G, Miyajima M, Bascones S, Mortha A, Puga I, Cassis L, Barra CM, Comerma L, Chudnovskiy A, Gentile M et al (2014) Innate lymphoid cells integrate stromal and immunological signals to enhance antibody production by splenic marginal zone B cells. Nat Immunol 15:354–364

Mandler R, Finkelman FD, Levine AD, Snapper CM (1993) IL-4 induction of IgE class switching by lipopolysaccharide-activated murine B cells occurs predominantly through sequential switching. J Immunol 150:407–418

Mao CS, Stavnezer J (2001) Differential regulation of mouse germline Ig γ1 and ε promoters by IL-4 and CD40. J Immunol 167:1522–1534

Meyer-Hermann M, Mohr E, Pelletier N, Zhang Y, Victora GD, Toellner KM (2012) A theory of germinal center B cell selection, division, and exit. Cell Rep 2:162–174

Mills FC, Thyphronitis G, Finkelman FD, Max EE (1992) Ig mu-epsilon isotype switch in IL-4-treated human B lymphoblastoid cells. Evidence for a sequential switch. J Immunol 149:1075–1085

Mills FC, Mitchell MP, Harindranath N, Max EE (1995) Human Ig S gamma regions and their participation in sequential switching to IgE. J Immunol 155:3021–3036

Mohrs K, Wakil AE, Killeen N, Locksley RM, Mohrs M (2005) A two-step process for cytokine production revealed by IL-4 dual-reporter mice. Immunity 23:419–429

Muramatsu M, Kinoshita K, Fagarasan S, Yamada S, Shinkai Y, Honjo T (2000) Class switch recombination and hypermutation require activation-induced cytidine deaminase (AID), a potential RNA editing enzyme. Cell 102:553–563

Nojima T, Haniuda K, Moutai T, Matsudaira M, Mizokawa S, Shiratori I, Azuma T, Kitamura D (2011) In-vitro derived germinal centre B cells differentially generate memory B or plasma cells in vivo. Nat Commun 2:465

Oberndorfer I, Schmid D, Geisberger R, Achatz-Straussberger G, Crameri R, Lamers M, Achatz G (2006) HS1-associated protein X-1 interacts with membrane-bound IgE: impact on receptor-mediated internalization. J Immunol 177:1139–1145

Ozaki K, Spolski R, Feng CG, Qi CF, Cheng J, Sher A, Morse HC 3rd, Liu C, Schwartzberg PL, Leonard WJ (2002) A critical role for IL-21 in regulating immunoglobulin production. Sci 298:1630–1634

Pesu M, Aittomaki S, Takaluoma K, Lagerstedt A, Silvennoinen O (2002) p38 Mitogen-activated protein kinase regulates interleukin-4-induced gene expression by stimulating STAT6-mediated transcription. J Biol Chem 277:38254–38261

Poggianella M, Bestagno M, Burrone OR (2006) The extracellular membrane-proximal domain of human membrane IgE controls apoptotic signaling of the B cell receptor in the mature B cell line A20. J Immunol 177:3597–3605

Punnonen J, Aversa G, Cocks BG, McKenzie AN, Menon S, Zurawski G, de Waal Malefyt R, de Vries JE (1993) Interleukin 13 induces interleukin 4-independent IgG4 and IgE synthesis and CD23 expression by human B cells. Proc Natl Acad Sci USA 90:3730–3734

Reinhardt RL, Liang HE, Locksley RM (2009) Cytokine-secreting follicular T cells shape the antibody repertoire. Nat Immunol 10:385–393

Rothman PB (2010) The transcriptional regulator NFIL3 controls IgE production. Trans Am Clin Climatol Assoc 121:156–171 discussion 171

Rothman P, Lutzker S, Cook W, Coffman R, Alt FW (1988) Mitogen plus interleukin 4 induction of C epsilon transcripts in B lymphoid cells. J Exp Med 168:2385–2389

Siebenkotten G, Esser C, Wabl M, Radbruch A (1992) The murine IgG1/IgE class switch program. Eur J Immunol 22:1827–1834

Smurthwaite L, Walker SN, Wilson DR, Birch DS, Merrett TG, Durham SR, Gould HJ (2001) Persistent IgE synthesis in the nasal mucosa of hay fever patients. Eur J Immunol 31:3422–3431

Snapper CM, Finkelman FD, Stefany D, Conrad DH, Paul WE (1988) IL-4 induces co-expression of intrinsic membrane IgG1 and IgE by murine B cells stimulated with lipopolysaccharide. J Immunol 141:489–498

Sugai M, Gonda H, Kusunoki T, Katakai T, Yokota Y, Shimizu A (2003) Essential role of Id2 in negative regulation of IgE class switching. Nat Immunol 4:25–30

Suto A, Nakajima H, Hirose K, Suzuki K, Kagami S, Seto Y, Hoshimoto A, Saito Y, Foster DC, Iwamoto I (2002) Interleukin 21 prevents antigen-induced IgE production by inhibiting germ line Cε transcription of IL-4-stimulated B cells. Blood 100:4565–4573

Takhar P, Smurthwaite L, Coker HA, Fear DJ, Banfield GK, Carr VA, Durham SR, Gould HJ (2005) Allergen drives class switching to IgE in the nasal mucosa in allergic rhinitis. J Immunol 174:5024–5032

Takhar P, Corrigan CJ, Smurthwaite L, O'Connor BJ, Durham SR, Lee TH, Gould HJ (2007) Class switch recombination to IgE in the bronchial mucosa of atopic and nonatopic patients with asthma. J Allergy Clin Immunol 119:213–218

Talay O, Yan D, Brightbill HD, Straney EE, Zhou M, Ladi E, Lee WP, Egen JG, Austin CD, Xu M et al (2012) IgE+ memory B cells and plasma cells generated through a germinal-center pathway. Nat Immunol 13:396–404

Talay O, Yan D, Brightbill HD, Straney EE, Zhou M, Ladi E, Lee WP, Egen JG, Austin CD, Xu M et al (2013) Addendum: IgE+ memory B cells and plasma cells generated through a germinal-center pathway. Nat Immunol 14:1302–1304

Tarlinton DM (2008) Evolution in miniature: selection, survival and distribution of antigen reactive cells in the germinal centre. Immunol Cell Biol 86:133–138

Turqueti-Neves A, Otte M, Prazeres da Costa O, Hopken UE, Lipp M, Buch T, Voehringer D (2014) B-cell-intrinsic STAT6 signaling controls germinal center formation. Eur J Immunol 44:2130–2138

Venkitaraman AR, Williams GT, Dariavach P, Neuberger MS (1991) The B-cell antigen receptor of the five immunoglobulin classes. Nature 352:777–781

Victora GD, Nussenzweig MC (2012) Germinal centers. Annu Rev Immunol 30:429–457

Vieira P, Rajewsky K (1988) The half-lives of serum immunoglobulins in adult mice. Eur J Immunol 18:313–316

Vinuesa CG, Tangye SG, Moser B, Mackay CR (2005) Follicular B helper T cells in antibody responses and autoimmunity. Nat Rev Immunol 5:853–865

Wan T, Beavil RL, Fabiane SM, Beavil AJ, Sohi MK, Keown M, Young RJ, Henry AJ, Owens RJ, Gould HJ et al (2002) The crystal structure of IgE Fc reveals an asymmetrically bent conformation. Nat Immunol 3:681–686

Wu LC, Zarrin AA (2014) The production and regulation of IgE by the immune system. Nat Rev Immunol 14:247–259

Xiong H, Curotto de Lafaille MA, Lafaille JJ (2012a) What is unique about the IgE response? Adv Immunol 116:113–141

Xiong H, Dolpady J, Wabl M, Curotto de Lafaille MA, Lafaille JJ (2012b) Sequential class switching is required for the generation of high affinity IgE antibodies. J Exp Med 209:353–364

Yang Z, Sullivan BM, Allen CD (2012) Fluorescent in vivo detection reveals that IgE⁺ B Cells are restrained by an intrinsic cell fate predisposition. Immunity 36:857–872

Yang Z, Robinson MJ, Allen CD (2014) Regulatory constraints in the generation and differentiation of IgE-expressing B cells. Curr Opin Immunol 28:64–70

Yoshida K, Matsuoka M, Usuda S, Mori A, Ishizaka K, Sakano H (1990) Immunoglobulin switch circular DNA in the mouse infected with Nippostrongylus brasiliensis: evidence for successive class switching from mu to epsilon via gamma 1. Proc Natl Acad Sci USA 87:7829–7833

Zhang K, Zhang L, Zhu D, Bae D, Nel A, Saxon A (2002) CD40-mediated p38 mitogen-activated protein kinase activation is required for immunoglobulin class switch recombination to IgE. J Allergy Clin Immunol 110:421–428

Zhang TT, Okkenhaug K, Nashed BF, Puri KD, Knight ZA, Shokat KM, Vanhaesebroeck B, Marshall AJ (2008) Genetic or pharmaceutical blockade of p110δ phosphoinositide 3-kinase enhances IgE production. J Allergy Clin Immunol 122:811–819 e812

Zhang TT, Makondo KJ, Marshall AJ (2012) p110delta phosphoinositide 3-kinase represses IgE switch by potentiating BCL6 expression. J Immunol 188:3700–3708

Molecular Mechanisms of IgE Class Switch Recombination

Pei Tong and Duane R. Wesemann

Abstract Immunoglobulin (Ig) E is the most tightly regulated of all Ig heavy chain (IgH) isotypes and plays a key role in atopic disease. The gene encoding for IgH in mature B cells consists of a variable region exon—assembled from component gene segments via V(D)J recombination during early B cell development—upstream of a set of IgH constant region C_H exons. Upon activation by antigen in peripheral lymphoid organs, B cells can undergo IgH class switch recombination (CSR), a process in which the initially expressed IgH μ constant region exons (Cμ) are deleted and replaced by one of several sets of downstream C_H exons (e.g., Cγ, Cε, and Cα). Activation of the IL-4 receptor on B cells, together with other signals, can lead to the replacement of Cμ with Cε resulting in CSR to IgE through a series of molecular events involving irreversible remodeling of the IgH locus. Here, we discuss the molecular mechanisms of CSR and the unique features surrounding the generation of IgE-producing B cells.

Contents

1 Introduction	22
2 The IgH Locus	23
2.1 V(D)J Recombination and B Cell Development	23
2.2 S Regions and Unique Structure Features of Sε	24
3 Mechanism of CSR	24
3.1 I Region Germline Transcription in CSR	25
3.2 Role of S Region Structure in AID Targeting	25
4 Unique Features of CSR to IgE	27
4.1 Transcriptional Control of Iε	27
4.2 Dual Activation of IgG1 and IgE	30
References	32

P. Tong · D.R. Wesemann (✉)
Department of Medicine, Division of Rheumatology, Immunology and Allergy,
Brigham and Women's Hospital and Harvard Medical School, Boston, MA 02115, USA
e-mail: dwesemann@research.bwh.harvard.edu

© Springer International Publishing Switzerland 2015
J.J. Lafaille and M.A. Curotto de Lafaille (eds.), *IgE Antibodies: Generation and Function*, Current Topics in Microbiology and Immunology 388,
DOI 10.1007/978-3-319-13725-4_2

1 Introduction

B cell immunoglobulin (Ig) E production is a tightly regulated mediator of allergic disease as well as host defense. As B cells develop, Ig heavy (IgH) and light (IgL) chain loci undergo V(D)J recombination and somatic hypermutation (SHM), which defines antibody specificity to antigen. The IgH locus can uniquely undergo an additional process that defines antibody function, namely IgH class switch recombination (CSR). CSR to IgE positions the variable region exon next to constant region exons that encode for IgE (Cε). The regulation of CSR to IgE involves transcriptional control, enzymatic modification of DNA leading to DNA double-stranded breaks (DSB), DNA repair processes, and permanent deletion of the previously expressed C exons (Wu and Zarrin 2014; Geha et al. 2003). In this review, we provide an overview of the intracellular events related to CSR with a focus on unique aspects of CSR to IgE (Table 1).

Table 1 Different regulation of transcription factors in Iε promoter and Iγ1 promoter

Transcription factor	Iε promoter	Iγ1 promoter	Species	Reference
NF-κB	+	+	Human and mouse	Delphin and Stavnezer (1995), Dryer and Covey (2005), Lin and Stavnezer (1996), Messner et al. (1997)
STAT6	+	+	Human and mouse	Stütz and Woisetschläger (1999), Linehan et al. (1998), Messner et al. (1997)
PU.1	+	?	Human	Stütz and Woisetschläger (1999)
Pax5	+	+	Human and mouse	Thienes et al. (1997), Max et al. (1995)
E2A	+	*	Mouse	Sugai et al. (2003)
NFIL3	+	*	Mouse	Kashiwada et al. (2010)
C/EBP β	+	−	Human and mouse	Mao and Stavnezer (2001), Shen and Stavnezer (2001), Lundgren et al. (1994)
Bcl6	−	−	Mouse	Harris et al. (1999, 2005)
Id2	−	*	Mouse	Sugai et al. (2003)
AP-1	+	*	Mouse	Shen and Stavnezer (2001)
SWAP-70	+	*	Mouse	Audzevich et al. (2013)

+ positive regulation; − negative regulation; * no effect; ? not clear

2 The IgH Locus

2.1 V(D)J Recombination and B Cell Development

The IgH locus in mice spans 2,300 kb on mouse chromosome 12 and 1,250 kb on human chromosome 14 (Lefranc 2001; Lefranc et al. 2005). The germline (unrearranged) configuration of the IgH locus consists of multiple V_H, D_H, and J_H gene segments located 5′ to several sets of C_H exons that correspond to the various IgH isotypes (Fig. 1). In developing progenitor (pro-) and precursor (pre-) B cells, transcription through V, (D), and J gene segments is accompanied by assembly of the IgH and IgL variable region exons through V(D)J recombination (Bassing et al. 2002), which is the primary driver of primary Ig diversification and provides an expansive repertoire of binding specificities. The RAG1 and RAG2 endonuclease complex (RAG) mediates V(D)J recombination by introducing DNA double-strand breaks within specific recombination signal sequences flanking germline V, (D), and J gene segments (Dudley et al. 2005). DNA end-joining factors complete the V(D)J recombination reaction by joining DNA ends together (Rooney et al. 2004; Lieber et al. 2003; Verkaik et al. 2002). Productive assembly of IgH and IgL V region exons leads to the surface IgM expression on immature B cells. This is followed by the initiation of alternative splicing events that result in the dual expression of both IgM and IgD on mature naïve B cells.

The C_H gene segments consist of several sets of exons each encoding an individual antibody constant region (Cμ for IgM, Cδ for IgD, Cγ for IgG, Cε for IgE, and Cα for IgA). As depicted in Fig. 1, Cμ is most proximal to the V region gene segments, while the most distal C regions, namely Cε and Cα, are over 100 kb downstream of Cμ in both mouse and human. Each C_H region, except Cδ, is a transcription unit equipped with a promoter and several downstream exons. The first intron of these C_H transcription units—located between the so-called I epsilon exon and downstream C_H exons—is called the switch (S) region and contains large stretches of repetitive DNA elements. As discussed in more detail below, activation

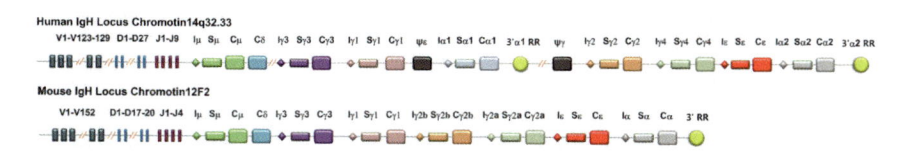

Fig. 1 Schematic presentation of IgH locus in human and mouse. The unassembled (germline) IgH variable region consists of V_H, D_H, and J_H clusters. The IgH constant region consists of several C_H gene segments. Each C_H gene segment, except for Cδ, is a noncoding transcription unit that includes an initial I exon (I), a large first intron termed the switch region (S), and a set of C_H exons (C) that contain open reading frames when in context of a transcript driven from a V region promoter, but not in the context of a transcript arising from the I region promoter. I promoters as well as 3′ regulatory region (3′RR in mouse and 3′α1 RR and 3′α2 RR in human) and other cis regulatory elements provide binding sites for transcription factors and other regulatory elements that affect transcription and IgH CSR. ψε and ψγ depict pseudogenes

of these C region transcription units, referred to as I region germline transcription, is required for CSR (Chaudhuri and Alt 2004).

Upon activation, mature naïve B cells can participate in further Ig diversification reactions including somatic hypermutation (SHM) and IgH class switch recombination (CSR). Both SHM and CSR are dependent upon the enzyme activation-induced cytidine deaminase (AID). SHM is a process wherein high-frequency point mutations occur within both IgH and IgL V region exons in germinal center (GC) B cells. These mutations result in altered binding Ig specificities that provide substrates for GC selection processes leading to Ig molecules with higher affinity to antigen in a process termed affinity maturation (Odegard and Schatz 2006; Shlomchik and Weisel 2012). IgH CSR replaces initially expressed Cμ exons with a set of downstream exons (Cγ, Cε, and Cα) leading to a switch from IgM to other IgH isotypes such as IgG, IgE, or IgA. As the IgH constant region isotype determines antibody function, IgH CSR positions the IgH V region of selected B cells in a different functional context.

2.2 S Regions and Unique Structure Features of Sε

CSR occurs between two S regions. In mouse, individual core sequences of S regions vary in size from 1 to 10 kb and are organized in the order 5'-VDJ-Sμ (4.0 kb), Sγ3 (2.5 kb), Sγ1 (8 kb), Sγ2b (5.0 kb), Sγ2a (2.5 kb), Sε (1.0 kb), and Sα (4.2 kb) (Zarrin et al. 2005). As discussed in more detail below, S regions are guanine rich on the non-template strand. Based on the type of repetitive sequences as well as homology in non-template strands, S regions can be classified into two groups. Sμ, Sε, and Sα contain 10–80 bp repeats and share motif GGGGT and GAGCT, while Sγ1, Sγ2a, Sγ2b, and Sγ3 contain 49–52 bp repeats and share motif AGCT which is also in Sμ, Sε, and Sα (Chaudhuri and Alt 2004). The AGCT motif is evolutionarily conserved in vertebrates such as amphibians, birds, and mammals and is a key target for the CSR machinery. In mammals, motifs containing G's in clusters of 2–5 together are required (Hackney et al. 2009). S region length and repetitiveness appear to influence IgH CSR efficiency in that more efficient S regions are longer in length and contain more repeats (Zarrin et al. 2005). In this regard, the Sε region is shortest in length and has the fewest amount of repetitive sequences compared to other S regions in mouse, making it the least efficient S region in general (Hackney et al. 2009).

3 Mechanism of CSR

AID initiates the CSR process through DNA deaminase activity. In this respect, AID deaminates deoxycytosine (dC) into deoxyuridine (dU) on a single-stranded DNA template. The resulting dC-to-dU mutation can then be processed via a number of pathways. The mutation may be repaired by a mismatch repair protein 2 (MSH2)-dependent process involving exonuclease activity and resynthesis of flanking DNA by an error-prone polymerase resulting in spreading of mutations from the original site of AID action. The mutation can also be replicated into a dC→dT mutation resulting in a point mutation at the site of AID action. Or the mutation may be processed into a DNA break through the action of uracil-N-glycosylase (UNG) and apyrimidinic endonuclease (APE1), which can remove the dU:dG mismatch resulting in a single-stranded DNA break. Double-stranded DNA breaks (DSBs) may result from multiple single-stranded DNA breaks in close proximity on both template and non-template DNA strands (Chaudhuri and Alt 2004; Chaudhuri et al. 2007). DSBs in both Sμ and a downstream S region—such as Sε in the case of CSR to IgE—can lead to deletion of the intervening DNA and joining of the Sμ and Sε breaks together through classical nonhomologous end-joining (C-NHEJ), or alternative end-joining (A-EJ) repair pathways (Yan et al. 2007). In terms of IgE, the post-CSR hybrid Sμ/Sε junction is located within an intron located between the V region exon and newly positioned Cε exons, which is spliced out of transcripts emanating from the V region promoter to produce productive IgE message. Alternative splicing of constant region exons can generate a membrane-bound or secreted Ig.

3.1 I Region Germline Transcription in CSR

Early studies showed that induction of CSR to a downstream C_H region correlated closely with antecedent transcription of the corresponding C_H region. Transcription of the Cε region proceeds through Iε exon/Sε region/Cε exons and terminates at a poly A site. Iε region germline transcript (εGLT) is spliced, capped, and transported out of the nucleus, but does not encode for protein due to lack of an open reading frame (Rothman et al. 1990). The importance of I epsilon exon and transcription in CSR was implicated decades ago when it was demonstrated that removal of 5′ flanking sequences of Sγ1 region abolished CSR to IgG1 (Jung et al. 1993). CSR to IgG2b was also abrogated in B cells lacking the Iγ2b promoter and I exon in spite of an intact Sγ2b region (Zhang et al. 1993). Replacement of Iγ2b with a strong transcriptional promoter (PKG-neo^r) was able to rescue CSR to IgG2b in Iγ2b-deficient cells (Seidl et al. 1998). Together, these results show that transcription is required for CSR.

3.2 Role of S Region Structure in AID Targeting

AID was discovered over a decade ago as the key mediator of both SHM and CSR (Muramatsu et al. 2000). AID is highly conserved in mouse and human with 92 % homology in sequence (Muto et al. 2000). Ectopic expression of AID in fibroblasts can induce CSR in a transcribed artificial switch construct reaching levels that parallel CSR in stimulated B cells, and the CSR is completely ablated in AID-deficient cells, indicating that AID is key for CSR (Muramatsu et al. 2000; Okazaki et al. 2002). Patients with mutations in AID also show a lack of IgH isotypes other than IgM (Revy et al. 2000).

In vitro observations have shown that AID deaminates dC into dU in single-stranded DNA (ssDNA) but not double-stranded DNA (dsDNA) or RNA (Ramiro et al. 2003; Chaudhuri et al. 2003; Dickerson et al. 2003; Pham et al. 2003) and AID is involved in stabilizing target sequences through the recruitment of replication protein A (RPA) (Chaudhuri et al. 2004; Basu et al. 2005). When AID is phosphorylated on serine 38, this induces the recruitment of RPA to transcribed S regions and enhances the activity of AID by virtue of its ability to stabilize single-stranded target DNA (Vuong et al. 2013). Experimental findings and current models discussed below account for AID acting on single-stranded DNA in the context of transcription.

Clustering of G's on the non-template strand in S regions can lead to RNA-DNA hybrids that are formed during transcription through S regions (Roy et al. 2008). Newly transcribed S region RNA is G rich and pairs with the C-rich template DNA with greater stability compared to its DNA complement, thus leaving a portion of the non-template G-rich DNA single stranded. This single-stranded loop of G-rich DNA is referred to an R loop and provides a model for how AID can target the single-stranded non-template DNA strand (Tracy et al. 2000; Yu et al. 2003; Roy et al. 2008). Substitution of only 5 % G with C by changing GGGGG into GGG and GGGG into GGG—while maintaining GC density—reduced the frequency of R loops dramatically from 6.7 to 0.23 %. When continuous G's changed into GG or G, no R loops were detected (Roy et al. 2008).

R loops have been detected in both Sμ and downstream S regions in vivo, including kilobase-sized R loops in Sγ3 and Sγ2b regions (Huang et al. 2007; Yu et al. 2003). Replacement of the endogenous Sγ1 region with an otherwise identical, but inverted sequence resulted in a 4-fold reduction of CSR to IgG1. Endogenous Sγ1 region is G rich on the non-template strand, which can form R loops in physiological orientation due to the favorable energetics of G-rich RNA interacting with C-rich DNA. Inverted S regions, which are C rich on the non-template strand, do not exhibit these characteristics and thus do not form detectable R loops (Shinkura et al. 2003). Replacement of Sγ1 region with *Xenopus* Sμ region, which is AT rich and similarly does not form R loops, results in a 4-fold decrease of CSR (Zarrin et al. 2004). These results suggest that although not required, R loops magnify the efficiency of CSR.

Although S regions in general are required for CSR, there appears to be flexibility regarding the requirements for specific S region sequences. In this context, substitution of Sα sequences with Sε and Sγ sequences did not change target specificity of CSR (Kinoshita et al. 1998). As mentioned above, while inverted Sγ1 and amphibian Sμ were still permissible for CSR, substitution of Sμ with vertebrate telomere sequences abolishes CSR completely (Junko et al. 2001). Inverted Sγ1 and vertebrate telomere sequences are both G rich, but the telomere sequences do not contain palindromic sequences, whereas both inverted Sγ1 and amphibian Sμ do.

As palindromic sequences can form secondary structures, such as stem loops, these findings may suggest that some form of secondary structure within S regions may be required. Other DNA structures are also reported in S region DNA sequence such as G quartets, in which guanines from four DNA strands are stabilized by G–G Hoogsteen Bondings and form a G4 planar structure (Dempsey et al. 1999). More complex structures involving AID and other factors during transcription have also been described (Yu and Lieber 2003; Shinkura et al. 2003). Although the roles of primary, secondary, and more complex tertiary structures remain to be fully elucidated, complex S region DNA structures are thought to lead to stalling of RNA polymerase II (RNA pol II)—to which AID is linked (Nambu et al. 2003)—thus providing CSR factors increased transcription-coupled S region access (Pavri et al. 2010; Xu et al. 2012; Kenter 2012).

Using an $MSH2^{-/-}$ $UNG^{-/-}$ genetic model system—in which AID-catalyzed deamination events can be tracked directly by C→T transitions—mutations are detected at equal frequencies in both template and non-template S region DNA (Xue et al. 2006), indicating that AID directly catalyzes deamination events on both template and non-template DNA. As template strand DNA does not form R loops, how AID targets the template DNA strand is an active area of research. In this regard, three models have been proposed to explain how AID may gain access to the template strand. One suggests that RNA polymerase generated DNA topological changes in S regions during transcription—where negatively supercoiled DNA is generated upstream of the active transcriptosome complex—provides AID targets to both DNA strands (Shen and Storb 2004; Besmer et al. 2006). Another model hypothesizes that RNase H leads to exposure of template DNA by degrading RNA transcripts in R loops (Huang et al. 2007; Lieber 2010). A third model implicates a role for the RNA exosome, which can degrade nascent RNA transcripts in R loops thus facilitating AID access to the template DNA strand (Basu et al. 2011; Nambu et al. 2003; Willmann et al. 2012; Pavri et al. 2010). In all of these models, non-template and template DNA strands in S regions must be accessible in the form of single-stranded DNA long enough for AID-mediated deamination to take place.

4 Unique Features of CSR to IgE

4.1 Transcriptional Control of Iε

As discussed above, accessibility of Sε to AID requires transcription of the germline Iε/Sε/Cε unit controlled by the Iε promoter upstream of Sε, as well as other cis regulatory elements, including a set of enhancer elements located at the 3′ end of the IgH locus termed the 3′ regulatory region (3′RR). Mouse IgH has one 3′RR downstream of the IgH C_H regions containing four enhancer segments, while human IgH has two 3′RR downstream of Cα1 and Cα2, respectively (Fig. 1). Each human 3′ RR contains enhancer segments homologous to mouse (Pinaud et al. 2011).

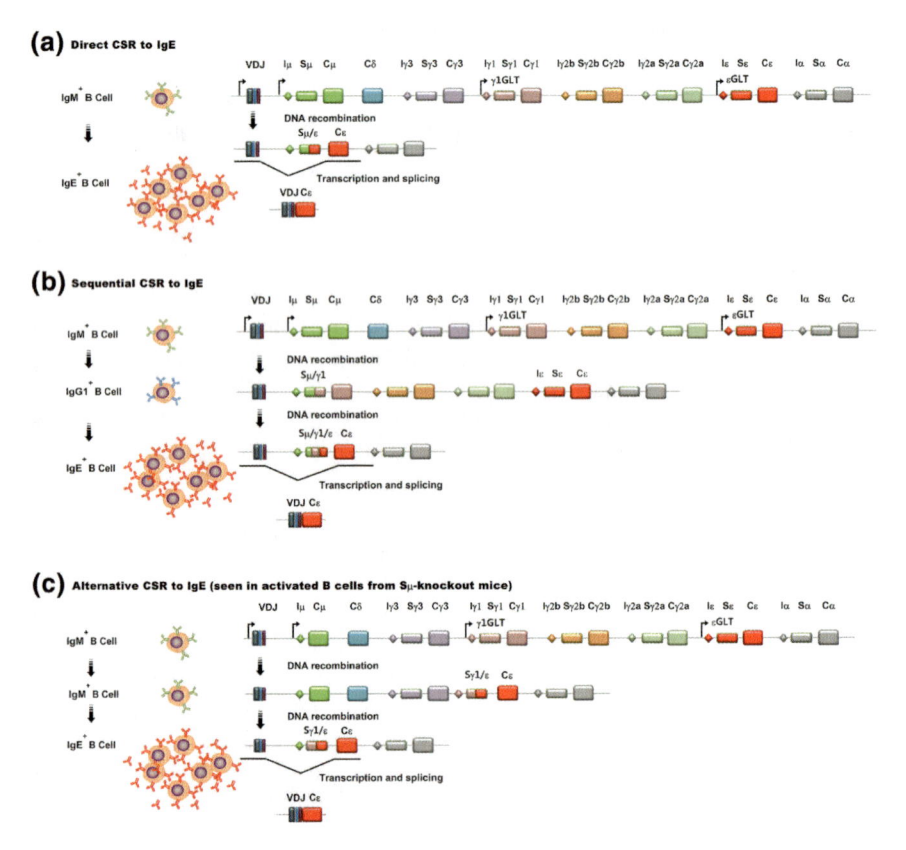

Fig. 2 Molecular pathways to IgE. **a–c** Schematic representations depicting the direct, sequential, and alternative class switch recombination. **a** One-step recombination of Sμ and Sε results in direct CSR to IgE. This pathway may predominate in immature B cells. **b** Recombination of Sμ region with Sγ1 region first causes the production of IgG1, then the hybrid Sμ/Sγ1 region recombines with Sε region leading to the production of IgE. In accordance with this process, B cell changes its surface immunoglobulin from IgM to IgG1, then to IgE with an IgG1⁺IgE⁺ intermediate. This pathway predominates in mature B cells. **c** An alternative class switch can be seen in Sμ-knockout mice, in which B cells undergo recombination of Sγ1 region with Sε region. Then, the hybrid Sγ1/Sε S region recombines with the weakened Sμ (Wu and Zarrin 2014; Zhang et al. 2010; Wesemann et al. 2011; Xiong et al. 2012)

Mice deficient in the entire IgH 3′ regulatory region (3′RR) show a 25-fold reduction of CSR to IgE and a more modest ~5 fold reduction in CSR to the other C_H isotypes (Vincent-Fabert et al. 2010) (Fig. 2).

Th2 cytokines, such as IL-4, are involved in driving CSR to IgE and IgG1. IL-4 binding to IL-4 receptor (IL-4R) activates signal transducer and activator of transcription 6 (STAT6). STAT6 is phosphorylated by JAK kinases, leading to homodimerization and translocation to the nucleus, where it can bind to the Iε promoter (Hebenstreit et al. 2006; Geha et al. 2003). STAT6 is coordinated on the Iε promoter with other inducible factors, such as NF-κB, which can be activated either through LPS stimulation or through T-cell interaction through CD40 signaling. The two transcription factors bind to adjacent areas on the Iε promoter and act synergistically to activation transcription (Shen and Stavnezer 1998).

The Iε promoter contains two binding sites for E2A proteins termed E boxes, which are both required to fully activate Iε transcription (Sugai et al. 2003). E2A proteins are negatively regulated by Id2, which is a helix-loop-helix transcription factor constitutively expressed in resting B cells (Ishiguro et al. 1995; Becker-Herman et al. 2002) and can be induced by transforming growth factor-β1 (TGF-β1) (Sugai et al. 2003). Id2$^{-/-}$ mice show an increased level of εGLT and elevated levels of IgE. Id2 negatively regulates Iε promoter by binding to activated transcription factor E2A, thus interfering with its binding to E boxes in Iε promoter. Although Id2 negatively regulates CSR to IgE, it does not appear to affect IgG1 (Sugai et al. 2003).

The zinc finger-containing transcriptional repressor B-cell lymphoma 6 (Bcl6) also plays an inhibitory role in IgE CSR (Harris et al. 2005). Bcl6-deficient mice undergo an increased CSR to IgE and IgG1 after IL-4 activation in vitro. There are Bcl6 binding sites within the Iε promoter and these were shown to overlap with the binding sites for STAT6 (Harris et al. 1999).

Several basic leucine zipper (bZIP) family members, such as ATF, AP-1, CCAAT/enhancer-binding protein (C/EBPβ), and IL-3-regulated (NFIL3, also called E4BP4), also bind to and can regulate the Iε promoter. After activation of the CD40 signaling pathway, AP-1 is rapidly upregulated and synergizes with STAT6 to activate the Iε promoter in mice. AP-1 recruits histone acetyl transferase, p300/CBP, which produces active histone modifications that may prolong the effect of AP-1 as it is induced quickly and transiently (Mao and Stavnezer 2001; Shen and Stavnezer 2001). In humans, C/EBPβ synergizes with STAT6, whereas mouse C/EBPβ may inhibit the synergism of AP-1 and STAT6 (Shen and Stavnezer 2001) and has been suggested to negatively regulate NF-κB on the Iγ1 promoter (Mao and Stavnezer 2001).

In response to stimulation with LPS plus IL-4, mouse B cells deficient in nuclear factor interleukin-3 regulated (NFIL3) greatly reduce the production of εGLT and CSR to IgE, while CSR to IgG1, IgM, IgG2a, IgG2b, and IgG3 are essentially unchanged (Kashiwada et al. 2010). Furthermore, unlike other bZIP transcriptional regulators, NFIL3 was found to bind to the Iε promoter but not other I promoters, indicating that in terms of C region regulation, NFIL3 may uniquely affect CSR to IgE (Kashiwada et al. 2010). Notably, NFIL3 has been shown to be involved in signaling circuits related to the circadian clock. The levels of NFIL3 are opposite to that of proline and acidic amino acid-rich (PAR) proteins, which are a group of transcription factors important in core circadian clockwork and fluctuate in a circadian fashion (Mitsui et al. 2001). Recently, the histone methyltransferase MLL3, which catalyzes active histone modification (H3K4me) in IgH locus I region promoters, was also found to be regulated by the circadian circuitry (Li et al. 2013; Valekunja et al. 2013). The protein levels of NFIL3 and MLL3 are both at their peak during the first half of day, suggesting that they may work together to add a circadian dimension to the transcriptional regulation of Iε and CSR to IgE. Such a dimension is an intriguing possibility that remains to be explored.

4.2 Dual Activation of IgG1 and IgE

IL-4 in combination with either CD40 activation or LPS stimulation induces murine B cell transcription of both the Iε and the Iγ1 promoters and thus may lead to switching to both IgG1 and IgE. Observations from over two decades ago that Sγ1 sequences can be found within Sμ/Sε junctions of both mouse and human IgE^{+} cells indicated that CSR can occur in a stepwise fashion. The first step is CSR from IgM to IgG, and the second step is a CSR reaction from IgG to IgE (Yoshida et al. 1990; Mills et al. 1992). It was later shown that B cells stimulated in vitro with IL-4 plus LPS, secreted IgG1 first, and then IgE after a delay of several hours. A similar delay also occured in the production of membrane-bound IgG1 (mIgG1) versus IgE (mIgE). A small number of intermediate cells containing both mIgG1 and mIgE can be detected, indicating B cell passage through a step where IgG1 protein is still present after having undergone CSR from Cγ1 to Cε at the DNA level with freshly produced IgE (Mandler et al. 1993). A similar pattern of IgG1 before IgE was observed in mature B cells upon stimulation with IL-4 plus a stimulating antibody against CD40 (anti-CD40) (Wesemann et al. 2011). The physiologic reasons for dual Iγ1 and Iε activation and the molecular coordination of CSR to IgG1 and IgE are areas of active research.

Although the frequency of sequential CSR to IgE through IgG1 is dominant over direct CSR to IgE in mature B cells (Jung et al. 1994), immature and transitional B cells have a preference to undergo CSR to IgE through a mechanism involving more direct IgM to IgE CSR (Wesemann et al. 2011). While the Iγ1 and Iε promoters are both induced substantially by IL-4 plus anti-CD40 in mature B cells, the preference for direct CSR from IgM to IgE in immature cells is explained molecularly by the relative preservation of Iε inducibility compared to the abrogated inducibility of the Iγ1 promoter in immature B cells (Wesemann et al. 2011). In addition, as young mice have a natural abundance of immature and transitional B cells in the periphery, splenic B cells from young mice demonstrate increased propensity to undergo CSR to IgE. The extent to which this mechanism contributes to the elevated IgE seen in some immune deficiencies as well as in youth compared to adults (Monroe et al. 1999; Melamed et al. 1998; Grundbacher 1976) remains to be uncovered.

Sμ contains the highest density of AID hotspots, and DSBs are more frequent in Sμ compared to any other downstream S region (Schrader et al. 2003, 2005). Although the evolutionary forces leading to Sμ as the most powerful S region are not fully known, we speculate that it may speak to the importance of simultaneous activation of two or more acceptor S regions. In the context of IL-4 plus anti-CD40 or LPS signaling, high DSB activity in the donor Sμ region appears to drive CSR to simultaneously activated Sγ1 and/or Sε. In this respect, the limiting DSB activity in the acceptor S regions may control CSR outcome by competing for "next best" status—in terms of transcriptional activation, AID targeting, and DSB activity—compared to Sμ. In the case of the simultaneously active triplex of Sμ, Sγ1, and Sε, excess DSB in Sμ may ensure that selection of the downstream acceptor S region resides within the respective relative strengths of Iγ1/Sγ1 and Iε/Sε to become transcriptionally active, recruit AID and form DSBs. If Sμ DSBs were not able to drive CSR in the setting of simultaneous activation of multiple S acceptor regions, the acceptor regions may undergo recombination themselves and thus reduce the flexibility and diversity of CSR outcomes. Such a scenario was observed in mice in which Sμ was weakened substantially through deletion of most of the core AID targets ($S\mu^{-/-}$ mice). $S\mu^{-/-}$ mice exhibited impaired CSR to IgG1, IgG2a, IgG2b, and IgG3 (Luby et al. 2001; Khamlichi et al. 2004). However, IL-4 plus anti-CD40-induced CSR to IgE is nearly the same levels in $S\mu^{-/-}$ mice as wild-type mice (Zhang et al. 2010) due to downstream Sγ1-to-Sε CSR occurring before involvement of the weakened Sμ. The Sγ1-to-Sε CSR results in deletion of the Cγ1 exons leaving a hybrid Sγ1/Sε S region as a second-step recombination partner with the inefficient, truncated Sμ sequence. Thus, as Cγ1 is located in between Cμ and Cε, the B cell looses the ability to undergo CSR to Cγ1 in $S\mu^{-/-}$ mice because Sγ1 and Sε recombination—and Cγ1 exon excision—will occur before either has a chance to recombine with the weakened Sμ, thus severely limiting the probability of Cγ1 as a CSR outcome due to its position along the IgH locus.

In addition to Sμ as the most powerful S region, at least three other regulatory mechanisms likely play a role in controlling CSR outcomes in the setting of simultaneous activation of Sμ, Sγ1, and Sε. Firstly, as discussed above, Sε is the weakest S region in terms of length, number of repeat elements, and AID target motifs, and Sγ1 is one of the strongest S regions given its large size. Hence, these primary structural DNA features would favor CSR from Sμ to Sγ1 first (unless Iγ1/Sγ1 cannot be transcriptionally activated as in the case of immature B cells). Evidence of this notion was derived from experiments wherein the Sε region was enhanced by replacing it with Sμ sequence. This resulted in more direct CSR to IgE in vitro and elevated IgE in vivo (Misaghi et al. 2013). Secondly, the Sγ1 region may play a role in actively repressing Sε accessibility to participate in CSR as deletion of Sγ1 results in increased levels of IgE in vitro and in vivo (Misaghi et al. 2010). However, deletion of the Iγ1 promoter did not affect the level of CSR to IgE (Jung et al. 1993, 1994). Thus, perhaps the Sγ1 region, but not the Iγ1 promoter, may be involved in regulating CSR to Cε. Thirdly, Sε appears to become accessible with delayed rate, even after Sμ/Sγ1 recombination has taken place. When activated, IgG1[+] B cells proceed from IgG1 to IgE with similar kinetics compared to IgM to IgE (Wesemann et al. 2012), indicating that regulatory mechanisms delay the timing of CSR to IgE.

Production of IgE through direct CSR or indirect CSR may differ in the extent to which affinity maturation may occur. In this regard, direct CSR to IgE tends to generate low-affinity IgE, while sequential CSR to IgE inherits V region somatic mutations and affinity selection that occurs during the IgG1 stage (Xiong et al. 2012).

After activation by IL-4, purified IgG1 cells derived from germinal centers and memory B cells underwent CSR to IgE, indicating that sequential CSR can be interrupted by an IgG1 stage that can be rounds of cellular proliferation and differentiation before switching to IgE (Erazo et al. 2007). So far, high-affinity IgE derived from memory IgM B cells has not been observed, and IgG1 deficient mice tend to produce low-affinity IgE indicating that B cells produce high-affinity IgE by way of IgG1 (Erazo et al. 2007; Xiong et al. 2012). Years to come will assuredly uncover a deeper molecular and cellular understanding of the developmental pathways that generate IgE.

Acknowledgements D.R.W. is supported by NIH grants AI089972 and AI113217, by the Mucosal Immunology Studies Team, and holds a Career Award for Medical Scientists from the Burroughs Wellcome Fund.

References

Audzevich T, Pearce G, Breucha M, Guenal G, Jessberger R (2013) Control of the STAT6-Bcl6 antagonism by SWAP-70 determines IgE production. J Immunol 190(10):4946–4955

Bassing CH, Swat W, Alt FW (2002) The mechanism and regulation of chromosomal V(D)J recombination. Cell 109:S45–S55

Basu U, Chaudhuri J, Alpert C, Dutt S, Ranganath S, Li G, Schrum JP, Manis JP, Alt FW (2005) The AID antibody diversification enzyme is regulated by protein kinase a phosphorylation. Nature 438(7067):508–511

Basu U, Meng FL, Keim C, Grinstein V, Pefanis E, Eccleston J, Zhang TT, Myers D, Wasserman CR, Wesemann DR, Januszyk K, Gregory RI, Deng HT, Lima CD, Alt FW (2011) The RNA exosome targets the AID cytidine deaminase to both strands of transcribed duplex DNA substrates. Cell 144(3):353–363

Becker-Herman S, Lantner F, Shachar I (2002) Id2 negatively regulates B cell differentiation in the spleen. J Immunol 168(11):5507–5513

Besmer E, Market E, Papavasiliou FN (2006) The transcription elongation complex directs activation-induced cytidine deaminase-mediated DNA deamination. Mol Cell Biol 26(11):4378–4385

Chaudhuri J, Alt FW (2004) Class-switch recombination: interplay of transcription, DNA deamination and DNA repair. Nat Rev Immunol 4(7):541–552

Chaudhuri J, Basu U, Zarrin A, Yan C, Franco S, Perlot T, Vuong B, Wang J, Phan RT, Datta A, Manis J, Alt FW (2007) Evolution of the immunoglobulin heavy chain class switch recombination mechanism. Adv Immunol 94:157–214

Chaudhuri J, Khuong C, Alt FW (2004) Replication protein a interacts with AID to promote deamination of somatic hypermutation targets. Nature 430(7003):992–998

Chaudhuri J, Tian M, Khuong C, Chua K, Pinaud E, Alt FW (2003) Transcription-targeted DNA deamination by the AID antibody diversification enzyme. Nature 422(6933):726–730

Delphin S, Stavnezer J (1995) Characterization of an interleukin 4 (IL-4) responsive region in the immunoglobulin heavy chain germline epsilon promoter: regulation by NF-IL-4, a C/EBP family member and NF-kappa B/p50. J Exp Med 181 (1):181–192

Dempsey LA, Sun H, Hanakahi LA, Maizels N (1999) G4 DNA binding by LR1 and its subunits, nucleolin and hnRNP D, a role for G-G pairing in immunoglobulin switch recombination. J Biol Chem 274(2):1066–1071

Dickerson SK, Market E, Besmer E, Papavasiliou FN (2003) AID mediates hypermutation by deaminating single stranded DNA. J Exp Med 197(10):1291–1296

Dryer RL, Covey LR (2005) A novel NF-kappa B-regulated site within the human I gamma 1 promoter requires p300 for optimal transcriptional activity. J Immunol 175(7):4499–4507

Dudley DD, Chaudhuri J, Bassing CH, Alt FW (2005) Mechanism and control of V(D)J recombination versus class switch recombination: similarities and differences. Adv Immunol 86:43–112

Erazo A, Kutchukhidze N, Leung M, Christ APG, Urban JF, de Lafaille MAC, Lafaille JJ (2007) Unique maturation program of the IgE response in vivo. Immunity 26(2):191–203

Geha RS, Jabara HH, Brodeur SR (2003) The regulation of immunoglobulin E class-switch recombination. Nat Rev Immunol 3(9):721–732

Grundbacher FJ (1976) Unusual elevation of ige levels during childhood. Experientia 32(8):1063–1064

Hackney JA, Misaghi S, Senger K, Garris C, Sun YL, Lorenzo MN, Zarrin AA (2009) DNA targets of AID: evolutionary link between antibody somatic hypermutation and class switch recombination. Adv Immunol 101:163–189

Harris MB, Chang CC, Berton MT, Danial KN, Zhang JD, Kuehner D, Ye BH, Kvatyuk M, Pandolfi PP, Cattoretti G, Dalla-Favera R, Rothman PB (1999) Transcriptional repression of Stat6-dependent interleukin-4-induced genes by Bcl-6: Specific regulation of I epsilon transcription and immunoglobulin E switching. Mol Cell Biol 19(10):7264–7275

Harris MB, Mostecki J, Rothman PB (2005) Repression of an interleukin-4-responsive promoter requires cooperative BCL-6 function. J Biol Chem 280(13):13114–13121

Hebenstreit D, Wirnsberger G, Horejs-Hoeck J, Duschl A (2006) Signaling mechanisms, interaction partners, and target genes of STAT6. Cytokine Growth Factor Rev 17(3):173–188

Huang FT, Yu KF, Balter BB, Selsing E, Oruc Z, Khamlichi AA, Hsieh CL, Lieber MR (2007) Sequence dependence of chromosomal R-loops at the immunoglobulin heavy-chain S mu class switch region. Mol Cell Biol 27(16):5921–5932

Ishiguro A, Spirin K, Shiohara M, Tobler A, Norton JD, Rigolet M, Shimbo T, Koeffler HP (1995) Expression of Id2 and Id3 mRNA in human lymphocytes. Leuk Res 19(12):989–996

Jung S, Rajewsky K, Radbruch A (1993) Shutdown of class switch recombination by deletion of a switch region control element. Science 259(5097):984–987

Jung S, Siebenkotten G, Radbruch A (1994) Frequency of immunoglobulin-E class switching is autonomously determined and independent of prior switching to other classes. J Exp Med 179 (6):2023–2026

Junko T, Kazuo K, Tasuku H (2001) Palindromic but not G-rich sequences are targets of class switch recombination. Int Immunol 13(4):495–505

Kashiwada M, Levy DM, McKeag L, Murray K, Schröder AJ, Canfield SM, Traver G, Rothman PB (2010) IL-4-induced transcription factor NFIL3/E4BP4 controls IgE class switching. P Natl Acad Sci USA 107(2):821–826

Kenter AL (2012) AID targeting is dependent on RNA polymerase II pausing. Semin Immunol 24 (4):281–286

Khamlichi AA, Glaudet F, Oruc Z, Denis V, Le Bert M, Cogné M (2004) Immunoglobulin class-switch recombination in mice devoid of any S mu tandem repeat. Blood 103(10):3828–3836

Kinoshita K, Tashiro J, Tomita S, Lee CG, Honjo T (1998) Target specificity of immunoglobulin class switch recombination is not determined by nucleotide sequences of S regions. Immunity 9 (6):849–858

Lefranc M-P, Clement O, Kaas Q, Duprat E, Chastellan P, Coelho I, Combres K, Ginestoux C, Giudicelli V, Chaume D, Lefranc G (2005) IMGT-choreography for immunogenetics and immunoinformatics. Silico Biol 5(1):45–60

Lefranc MP (2001) Nomenclature of the human immunoglobulin heavy (IGH) genes. Exp Clin Immunogenet 18(2):100–116

Li GD, White CA, Lam T, Pone EJ, Tran DC, Hayama KL, Zan H, Xu ZM, Casali P (2013) Combinatorial H3K9acS10ph histone modification in IgH locus S regions targets 14-3-3 adaptors and AID to specify antibody class-switch DNA recombination. Cell Rep 5 (3):702–714

Lieber MR (2010) The mechanism of double-strand DNA break repair by the nonhomologous DNA end-joining pathway. Annu Rev Biochem 79:181–211

Lieber MR, Ma YM, Pannicke U, Schwarz K (2003) Mechanism and regulation of human nonhomologous DNA end-joining. Nat Rev Mol Cell Biol 4(9):712–720

Lin SC, Stavnezer J (1996) Activation of NF-kappa B/Rel by CD40 engagement induces the mouse germline immunoglobulin C gamma 1 promoter. Mol Cell Biol 16(9):4591–4603

Linehan LA, Warren WD, Thompson PA, Grusby MJ, Berton MT (1998) STAT6 is required for IL-4-induced germline Ig gene transcription and switch recombination. J Immunol 161(1):302–310

Luby TM, Schrader CE, Stavnezer J, Selsing E (2001) The mu switch region tandem repeats are important, but not required, for antibody class switch recombination. J Exp Med 193(2):159–168

Lundgren M, Larsson C, Femino A, Xu MZ, Stavnezer J, Severinson E (1994) Activation of the Ig germ-line gamma-1 promoter—involvement of C/enhancer-binding protein transcription factors and their possible interaction with an NF-IL-4 site. J Immunol 153(7):2983–2995

Mandler R, Finkelman FD, Levine AD, Snapper CM (1993) IL-4 induction of IgE class switching by lipopolysaccharide-activated murine B cells occurs predominantly through sequential switching. J Immunol 150(2):407–418

Mao C, Stavnezer J (2001) Differential regulation of mouse germline Ig gamma 1 and epsilon promoters by IL-4 and CD40. J Immunol 167(3):1522–1534

Max EE, Wakatsuki Y, Neurath MF, Strober W (1995) The role of BSAP in immunoglobulin isotype switching and B cell proliferation. Curr Top Microbiol 194:449–458

Melamed D, Benschop RJ, Cambier JC, Nemazee D (1998) Developmental regulation of B lymphocyte immune tolerance compartmentalizes clonal selection from receptor selection. Cell 92(2):173–182

Messner B, Stütz AM, Albrecht B, Peiritsch S, Woisetschläger M (1997) Cooperation of binding sites for STAT6 and NF kappa B/rel in the IL-4-induced up-regulation of the human IgE germline promoter. J Immunol 159(7):3330–3337

Mills FC, Thyphronitis G, Finkelman FD, Max EE (1992) Ig mu-epsilon isotype switch in IL-4-treated human B-lymphoblastoid cells—evidence for a sequential switch. J Immunol 149(3):1075–1085

Misaghi S, Garris CS, Sun Y, Nguyen A, Zhang J, Sebrell A, Senger K, Yan D, Lorenzo MN, Heldens S, Lee WP, Xu M, Wu J, DeForge L, Sai T, Dixit VM, Zarrin AA (2010) Increased targeting of donor switch region and IgE in S gamma 1-deficient B Cells. J Immunol 185(1):166–173

Misaghi S, Senger K, Sai T, Qu Y, Sun YL, Hamidzadeh K, Nguyen A, Jin ZY, Zhou MJ, Yan DH, Lin WY, Lin ZH, Lorenzo MN, Sebrell A, Ding JB, Xu M, Caplazi P, Austin CD, Balazs M, Roose-Girma M, DeForge L, Warming S, Lee WP, Dixit VM, Zarrin AA (2013) Polyclonal hyper-IgE mouse model reveals mechanistic insights into antibody class switch recombination. Proc Natl Acad Sci USA 110(39):15770–15775

Mitsui S, Yamaguchi S, Matsuo T, Ishida Y, Okamura H (2001) Antagonistic role of E4BP4 and PAR proteins in the circadian oscillatory mechanism. Gene Dev 15(8):995–1006

Monroe RJ, Seidl KJ, Gaertner F, Han SH, Chen F, Sekiguchi J, Wang JY, Ferrini R, Davidson L, Kelsoe G, Alt FW (1999) RAG2: GFP knockin mice reveal novel aspects of RAG2 expression in primary and peripheral lymphoid tissues. Immunity 11(2):201–212

Muramatsu M, Kinoshita K, Fagarasan S, Yamada S, Shinkai Y, Honjo T (2000) Class switch recombination and hypermutation require activation-induced cytidine deaminase (AID), a potential RNA editing enzyme. Cell 102(5):553–563

Muto T, Muramatsu M, Taniwaki M, Kinoshita K, Honjo T (2000) Isolation, tissue distribution, and chromosomal localization of the human activation-induced cytidine deaminase (AID) gene. Genomics 68(1):85–88

Nambu Y, Sugai M, Gonda H, Lee CG, Katakai T, Agata Y, Yokota Y, Shimizu A (2003) Transcription-coupled events associating with immunoglobulin switch region chromatin. Science 302(5653):2137–2140

Odegard VH, Schatz DG (2006) Targeting of somatic hypermutation. Nat Rev Immunol 6(8):573–583

Okazaki IM, Kinoshita K, Muramatsu M, Yoshikawa K, Honjo T (2002) The AID enzyme induces class switch recombination in fibroblasts. Nature 416(6878):340–345

Pavri R, Gazumyan A, Jankovic M, Di Virgilio M, Klein I, Ansarah-Sobrinho C, Resch W, Yamane A, San-Martin BR, Barreto V, Nieland TJ, Root DE, Casellas R, Nussenzweig MC (2010) Activation-induced cytidine deaminase targets DNA at sites of RNA polymerase II stalling by interaction with Spt5. Cell 143(1):122–133

Pham P, Bransteitter R, Petruska J, Goodman MF (2003) Processive AID-catalysed cytosine deamination on single-stranded DNA simulates somatic hypermutation. Nature 424 (6944):103–107

Pinaud E, Marquet M, Fiancette R, Péron S, Vincent-Fabert C, Denizot Y, Cogné M (2011) The IgH locus 3' regulatory region: pulling the strings from behind. Adv Immunol 110:27–70

Ramiro AR, Stavropoulos P, Jankovic M, Nussenzweig MC (2003) Transcription enhances AID-mediated cytidine deamination by exposing single-stranded DNA on the nontemplate strand. Nat Immunol 4(5):452–456

Revy P, Muto T, Levy Y, Geissmann F, Plebani A, Sanal O, Catalan N, Forveille M, Dufourcq-Lagelouse R, Gennery A, Tezcan I, Ersoy F, Kayserili H, Ugazio AG, Brousse N, Muramatsu M, Notarangelo LD, Kinoshita K, Honjo T, Fischer A, Durandy A (2000) Activation-induced cytidine deaminase (AID) deficiency causes the autosomal recessive form of the hyper-IgM syndrome (HIGM2). Cell 102(5):565–575

Rooney S, Chaudhuri J, Alt FW (2004) The role of the non-homologous end-joining pathway in lymphocyte development. Immunol Rev 200:115–131

Rothman P, Chen YY, Lutzker S, Li SC, Stewart V, Coffman R, Alt FW (1990) Structure and expression of germ line immunoglobulin heavy-chain epsilon-transcripts—interleukin-4 plus lipopolysaccharide-directed switching to C-epsilon. Mol Cell Biol 10(4):1672–1679

Roy D, Yu KF, Lieber MR (2008) Mechanism of R-loop formation at immunoglobulin class switch sequences. Mol Cell Biol 28(1):50–60

Schrader CE, Bradley SP, Vardo J, Mochegova SN, Flanagan E, Stavnezer J (2003) Mutations occur in the Ig Smu region but rarely in Sgamma regions prior to class switch recombination. EMBO J 22(21):5893–5903

Schrader CE, Linehan EK, Mochegova SN, Woodland RT, Stavnezer J (2005) Inducible DNA breaks in Ig S regions are dependent on AID and UNG. J Exp Med 202(4):561–568

Seidl KJ, Bottaro A, Vo A, Zhang J, Davidson L, Alt FW (1998) An expressed neo(r) cassette provides required functions of the I(gamma)2b exon for class switching. Int Immunol 10 (11):1683–1692

Shen CH, Stavnezer J (1998) Interaction of Stat6 and NF-kappa B: direct association and synergistic activation of interleukin-4-induced transcription. Mol Cell Biol 18(6):3395–3404

Shen CH, Stavnezer J (2001) Activation of the mouse Ig germline epsilon promoter by IL-4 is dependent on AP-1 transcription factors. J Immunol 166(1):411–423

Shen HM, Storb U (2004) Activation-induced cytidine deaminase (AID) can target both DNA strands when the DNA is supercoiled. Proc Natl Acad Sci USA 101(35):12997–13002

Shinkura R, Tian M, Smith M, Chua K, Fujiwara Y, Alt FW (2003) The influence of transcriptional orientation on endogenous switch region function. Nat Immunol 4(5):435–441

Shlomchik MJ, Weisel F (2012) Germinal centers. Immunol Rev 247:5–10

Stützl AM, Woisetschläger M (1999) Functional synergism of STAT6 with either NF-kappa B or PU.1 to mediate IL-4-induced activation of IgE germline gene transcription. J Immunol 163 (8):4383–4391

Sugai M, Gonda H, Kusunoki T, Katakai T, Yokota Y, Shimizu A (2003) Essential role of Id2 in negative regulation of IgE class switching. Nat Immunol 4(1):25–30

Thienes CP, DeMonte L, Monticelli S, Busslinger M, Gould HJ, Vercelli D (1997) The transcription factor B cell-specific activator protein (BSAP) enhances both IL-4- and CD40-mediated activation of the human epsilon germline promoter. J Immunol 158(12):5874–5882

Tracy RB, Hsieh CL, Lieber MR (2000) Stable RNA/DNA hybrids in the mammalian genome: inducible intermediates in immunoglobulin class switch recombination. Science 288 (5468):1058–1061

Valekunja UK, Edgar RS, Oklejewicz M, van der Horst GTJ, O'Neill JS, Tamanini F, Turner DJ, Reddy AB (2013) Histone methyltransferase MLL3 contributes to genome-scale circadian transcription. Proc Natl Acad Sci USA 110(4):1554–1559

Verkaik NS, Esveldt-van Lange REE, van Heemst D, Brüggenwirth HT, Hoeijmakers JHJ, Zdzienicka MZ, van Gent DC (2002) Different types of V(D)J recombination and end-joining defects in DNA double-strand break repair mutant mammalian cells. Eur J Immunol 32 (3):701–709

Vincent-Fabert C, Fiancette R, Pinaud E, Truffinet V, Cogné N, Cogné M, Denizot Y (2010) Genomic deletion of the whole IgH 3′ regulatory region (hs3a, hs1,2, hs3b, and hs4) dramatically affects class switch recombination and Ig secretion to all isotypes. Blood 116 (11):1895–1898

Vuong BQ, Herrick-Reynolds K, Vaidyanathan B, Pucella JN, Ucher AJ, Donghia NM, Gu XW, Nicolas L, Nowak U, Rahman N, Strout MP, Mills KD, Stavnezer J, Chaudhuri J (2013) A DNA break- and phosphorylation-dependent positive feedback loop promotes immunoglobulin class-switch recombination. Nat Immunol 14(11):1183–U1107

Wesemann DR, Magee JM, Boboila C, Calado DP, Gallagher MP, Portuguese AJ, Manis JP, Zhou X, Recher M, Rajewsky K, Notarangelo LD, Alt FW (2011) Immature B cells preferentially switch to IgE with increased direct Smu to Sepsilon recombination. J Exp Med 208(13):2733–2746

Wesemann DR, Portuguese AJ, Magee JM, Gallagher MP, Zhou X, Panchakshari RA, Alt FW (2012) Reprogramming IgH isotype-switched B cells to functional-grade induced pluripotent stem cells. Proc Natl Acad Sci USA 109(34):13745–13750

Willmann KL, Milosevic S, Pauklin S, Schmitz KM, Rangam G, Simon MT, Maslen S, Skehel M, Robert I, Heyer V, Schiavo E, Reina-San-Martin B, Petersen-Mahrt SK (2012) A role for the RNA pol II-associated PAF complex in AID-induced immune diversification. J Exp Med 209 (11):2099–2111

Wu LC, Zarrin AA (2014) The production and regulation of IgE by the immune system. Nat Rev Immunol 14(4):247–259

Xiong HZ, Dolpady J, Wabl M, de Lafaille MAC, Lafaille JJ (2012) Sequential class switching is required for the generation of high affinity IgE antibodies. J Exp Med 209(2):353–364

Xu Z, Zan H, Pone EJ, Mai T, Casali P (2012) Immunoglobulin class-switch DNA recombination: induction, targeting and beyond. Nat Rev Immunol 12(7):517–531

Xue KM, Rada C, Neuberger MS (2006) The in vivo pattern of AID targeting to immunoglobulin switch regions deduced from mutation spectra in msh2(−/−) ung(−/−) mice. J Exp Med 203 (9):2085–2094

Yan CT, Boboila C, Souza EK, Franco S, Hickernell TR, Murphy M, Gumaste S, Geyer M, Zarrin AA, Manis JP, Rajewsky K, Alt FW (2007) IgH class switching and translocations use a robust non-classical end-joining pathway. Nature 449(7161):478–U479

Yoshida K, Matsuoka M, Usuda S, Mori A, Ishizaka K, Sakano H (1990) Immunoglobulin switch circular DNA in the mouse infected with nippostrongylus-brasiliensis—evidence for successive class switching from mu to epsilon via gamma-1. Proc Natl Acad Sci USA 87(20):7829–7833

Yu KF, Chedin F, Hsieh CL, Wilson TE, Lieber MR (2003) R-loops at immunoglobulin class switch regions in the chromosomes of stimulated B cells. Nat Immunol 4(5):442–451

Yu KF, Lieber MR (2003) Nucleic acid structures and enzymes in the immunoglobulin class switch recombination mechanism. DNA Repair 2(11):1163–1174

Zarrin AA, Alt FW, Chaudhuri J, Stokes N, Kaushal D, Du Pasquier L, Tian M (2004) An evolutionarily conserved target motif for immunoglobulin class-switch recombination. Nat Immunol 5(12):1275–1281

Zarrin AA, Tian M, Wang J, Borjeson T, Alt FW (2005) Influence of switch region length on immunoglobulin class switch recombination. Proc Natl Acad Sci USA 102(7):2466–2470

Zhang J, Bottaro A, Li S, Stewart V, Alt FW (1993) A selective defect in IgG2b switching as a result of targeted mutation of the I-gamma-2b promoter and exon. EMBO J 12(9):3529–3537
Zhang T, Franklin A, Boboila C, McQuay A, Gallagher MP, Manis JP, Khamlichi AA, Alt FW (2010) Downstream class switching leads to IgE antibody production by B lymphocytes lacking IgM switch regions. Proc Natl Acad Sci USA 107(7):3040–3045

Anti-IgE Therapy: Clinical Utility and Mechanistic Insights

Stephanie L. Logsdon and Hans C. Oettgen

Abstract As the major trigger of acute allergic reactions, IgE has long been considered an ideal target for anti-allergy treatments. Omalizumab, first approved by the USA in 2003 and now in use in many other countries is a humanized monoclonal antibody that binds serum IgE. Anti-IgE therapy using omalizumab reduces circulating free IgE levels and blocks both early and late-phase reactions to allergen challenge. It has proven effective for allergic asthma and is currently being evaluated for use in a number of other atopic conditions including allergic rhinitis and chronic idiopathic urticaria. Clinical observations and mechanistic studies with omalizumab have shed new light on the multifaceted roles of IgE in immune homeostasis and in allergic disease.

Contents

1	IgE Functions in Immune Homeostasis and Pathology	40
	1.1 IgE, IgE Receptor, and Effector Cells	40
	1.2 IgE and Regulation of IgE Receptors	43
	1.3 Mast Cell Homeostasis	43
	1.4 Antiviral Responses	44
2	Omalizumab: Structure and Pharmacology	45
	2.1 Structure: Humanized MAb Against IgE	45
	2.2 IgE–Omalizumab Interaction	46
	2.3 Pharmacokinetics	46
3	Early Clinical Studies in Asthma	47
	3.1 Inhalation Challenge	47
	3.2 Early and Late-phase Suppression, Further Analyses	47
4	Asthma Studies	47
5	Pediatric Asthma	48

S.L. Logsdon · H.C. Oettgen (✉)
Division of Immunology, Boston Children's Hospital, Harvard Medical School, Boston, MA 02115, USA
e-mail: hans.oettgen@childrens.harvard.edu

© Springer International Publishing Switzerland 2015
J.J. Lafaille and M.A. Curotto de Lafaille (eds.), *IgE Antibodies: Generation and Function*, Current Topics in Microbiology and Immunology 388,
DOI 10.1007/978-3-319-13725-4_3

6	Omalizumab in Other Atopic Diseases	49
	6.1 Allergic Rhinitis	50
	6.2 Rush Immunotherapy	50
	6.3 Food-induced Anaphylaxis and Oral Immunotherapy	51
	6.4 Chronic Idiopathic Urticaria	52
	6.5 Atopic Dermatitis	53
	6.6 Eosinophilic Gastrointestinal Disorders	54
7	Mechanisms	54
	7.1 Blockade of Early and Late-phase Hypersensitivity	54
	7.2 Regulation of IgE Receptors	55
	7.3 Effects on Viral Immunity	56
8	Future Directions	56
References		57

1 IgE Functions in Immune Homeostasis and Pathology

1.1 IgE, IgE Receptor, and Effector Cells

Atopic diseases are characterized by production of allergen-specific IgE antibodies. These arise when IgM^+ or IgG^+ B cells of committed antigenic specificity are driven into the process of class switch recombination (CSR) by T helper-2 (Th2) cells. As with CSR to other isotypes, the IgE switch can occur in germinal centers of secondary lymphoid tissues. This may be where some or all IgE memory resides and where affinity maturation occurs during the stage of pre-IgE-switched IgG^+ B cell intermediates (Xiong et al. 2012). However, CSR to IgE is not restricted to lymphoid tissues; it also occurs directly in mucosal tissue sites of allergen exposure (Cameron et al. 2003).

Several unique features of the IgE antibodies bear on the design and implementation of effective strategies of IgE blockade. IgE is the least prevalent antibody in circulation, with concentrations of 50–100 ng/ml in normal individuals, logs lower than the normal level of IgG (5–10 mg/ml). IgE also has a short circulating half-life of only 2–3 days, but it is stable in tissues for many weeks where is it tightly bound to FcεRI on mast cells and not readily accessible to inhibitors designed to target-free IgE. This latter aspect of IgE physiology poses one of the major challenges to the designers of novel anti-IgE strategies (Kubo et al. 2003; Gould et al. 2003). The creation of an agent which would displace tightly bound IgE antibodies from FcεRI would be the ultimate aspiration for a designer of anti-IgE therapeutics.

IgE shares a similar molecular structure with other immunoglobulin isotypes, with two pairs of identical heavy and light chains. However, the constant regions of the ε-heavy chains are comprised of four Cε domains, Cε1–Cε4, in contrast to three Cγ domains in IgG proteins. The constant regions determine the isotype-specific functions of IgE, including binding to its high- and low-affinity receptors. Membrane-bound IgE expressed on the surface of IgE^+ B cells is the product of a splice isoform containing hydrophobic sequences encoded by M1 and M2 exons (Gould et al. 2003). Monoclonal antibodies targeting the unique M1 domain of transmembrane IgE are under development as potential inhibitors of IgE production (Brightbill et al. 2010).

IgE antibodies exert their biological effects via several receptors. The two major ones are commonly referred to as the "high-affinity receptor," FcεRI, and the "low-affinity receptor," CD23, the latter having a more respectable affinity for IgE than its name would imply. FcεRI is expressed on mast cells and basophils as a $\alpha\beta\gamma_2$ tetramer (Fig. 1). A trimeric $\alpha\gamma_2$ form of FcεRI exists on the surface of dendritic cells, Langerhans cells, and eosinophils in humans. In rodents, FcεRI expression is generally restricted to mast cells and basophils, and trimeric FcεRI expression is observed only under limited circumstances such as viral infection (Dombrowicz et al. 2000; Grayson et al. 2007; Kraft and Kinet 2007; Lin et al. 1996). The α subunit is a type I integral membrane protein which binds the Fc region of IgE. It is comprised of two extracellular immunoglobulin domains, which

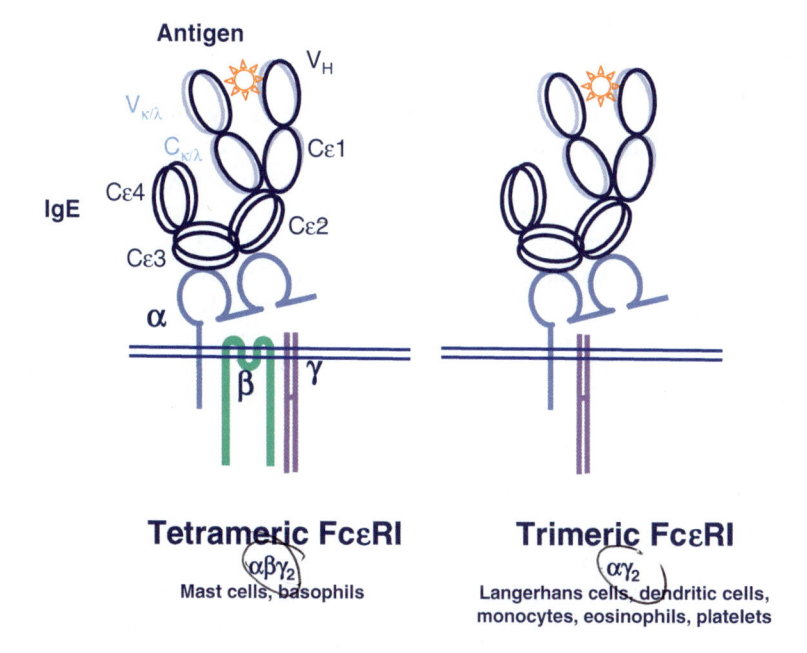

Fig. 1 Structure of the high-affinity IgE receptor, FcεRI. FcεRI is assembled in a tetrameric $\alpha\beta\gamma_2$ form on mast cells and basophils and a trimeric $\alpha\gamma_2$ form on numerous other cell types. The α subunit is a type I integral membrane protein which binds the $C\varepsilon_{2-3}$ domains of the Fc region of IgE. It is comprised of two extracellular immunoglobulin domains, a transmembrane domain and a short intracellular tail. The β-chain, a member of the tetraspanin family of proteins that cross the plasma membrane four times, serves to amplify receptor signal transduction. A pair of disulfide-linked γ-chains, also shared with FcγRIII, contains several intracellular ITAM motifs and is primarily responsible for docking of the signaling protein tyrosine kinase, SYK, in the initiation of IgE-triggered signaling

bind the Cε_{2-3} regions, a transmembrane domain and a short intracellular tail. In comparison with FcγRs, FcεRI is typically completely saturated at normal physiological concentrations of IgE due to the exceedingly low K_d (~ 1 nm) for IgE binding. This underlies the persistence of allergen-specific IgE in tissues for many weeks even in individuals being treated with anti-IgE.

Neither the β subunit nor γ subunits assist in ligand binding. The β chain provides accessory signaling and amplification, while the γ chain, which is shared with FcγRIII, is the main activator of key signaling pathways (Kraft and Kinet 2007). Cross-linking of FcεRI causes signal transduction by phosphorylation of tyrosine-based activation motifs (ITAM) in the γ subunit through activation of constitutively receptor-associated tyrosine kinases.

The low-affinity IgE receptor, CD23, is completely distinct from FcεRI. It is a type II transmembrane glycoprotein with a c-terminal extracellular globular domain containing a calcium-dependent lectin domain. This is followed by an alpha-helical coiled-coil stalk region with a short n-terminal cytoplasmic domain (Acharya et al. 2010). The stalk region contains leucine zipper motifs that mediate oligomerization of CD23. Trimeric CD23 exhibits an avidity for IgE of 10–100 nM which approaches the affinity of FcεRI for IgE (Gould and Sutton 2008). The membrane-bound form of CD23 is susceptible to proteolysis, generating a free receptor known as soluble CD23 that retains the capacity to bind IgE.

IgE antibodies are best known for their role in inducing immediate hypersensitivity reactions, in which exposure to an allergen results in rapid development of allergic symptoms. These reactions are initiated by activation of the high-affinity receptor FcεRI on effector cells including mast cells and basophils. FcεRI aggregation induced by polyvalent antigen-induced IgE cross-linking incites the rapid release of preformed mediators along with de novo synthesis of lipid mediators, which act on target tissues to trigger the early phase allergic response within minutes of allergen exposure. Hours later FcεRI-activated cells produce proinflammatory and immunomodulating cytokines and chemokines that promote recruitment of cells including eosinophils, monocytes, and T cells (Sutton and Gould 1993). These events can elicit "late-phase" symptoms of allergy, in which a second wave of allergic symptoms is experienced 8–12 h after the acute reaction. These proinflammatory mediators, when produced chronically in the course of recurrent IgE-triggered reactions also drive the chronic allergic inflammation characteristic of a number of atopic diseases. In the setting of asthma, mediators released by degranulation of mast cells and basophils act on endothelial cells to increase vascular permeability leading to edema of the airway wall. Mucus production and constriction of airway smooth muscles, induced by the same mediators, lead to further airflow airway narrowing. In an individual with aeroallergen sensitivity, these events manifest as acute airflow obstruction, measured in the clinical laboratory as a drop in forced expiratory volume of air in 1 s (FEV_1). Asthmatic individuals exhibit acute drops in FEV_1 immediately after allergen challenge and late phase decreases 8–12 h later. In some chronic asthmatics, cellular infiltration of the airway wall and chronic fibrotic changes cause stable airflow obstruction with a fixed decrease in FEV_1.

The major rationale underlying the development of anti-IgE therapies has been that blockade of acute and late-phase reactions precipitated by allergen exposure would reduce symptoms. There have additionally been expectations that interference with chronic recurrent acute reactions to allergen would diminish the pool of mast cell and basophil-derived mediators that might drive cellular recruitment and chronic allergic inflammation. However, recent findings regarding functions of IgE in the regulation of its own receptors and in mast cell homeostasis suggest additional mechanisms whereby IgE blockade might be beneficial.

1.2 IgE and Regulation of IgE Receptors

IgE plays an important role in the regulation and expression of both its high- and low-affinity receptors. Our studies with IgE-deficient mice revealed reduced expression of both FcεRI on their mast cells and basophils and CD23 on their B cells (Kisselgof and Oettgen 1998; Yamaguchi et al. 1997). Reconstitution of normal circulating IgE levels via i.v. infusion restored surface expression of both receptors to normal levels. In human studies, Saini et al. (2000) showed a direct correlation between total serum IgE levels and FcεRI cell surface expression, across a range of atopic and nonatopic conditions. IgE binding to CD23 inhibits cleavage of the receptor by proteases, thus allowing the IgE-CD23 complex to remain on the effector cell surface (MacGlashan 2008). Similarly, binding of IgE to cell surface FcεRI stabilizes the receptor and prevents both removal of the receptor from the surface of the effector cell, and degradation by proteases. Empty FcεRI receptors are rapidly internalized, and the presence of IgE in the milieu of effector cells supports accumulation of FcεRI at the cell membrane (Borkowski et al. 2001; Saini et al. 2000). It is very likely that some of these effects of IgE on the stability and expression of its receptors underlie the benefits of anti-IgE therapy.

1.3 Mast Cell Homeostasis

Up until recently, the accepted paradigm of FcεRI-mediated IgE signaling held that cross-linking of receptor-bound IgE by polyvalent allergen and consequent aggregation of FcεRI in the membrane were critical for signaling. A mast cell or basophil bearing FcεRI-bound IgE was considered a loaded gun, ready for allergen-driven triggering but unaffected by the mere presence of IgE. A number of studies now reveal that IgE itself, in the absence of exogenous allergen, can exert significant effects. These findings have significance with respect to consideration of potential mechanisms of action of IgE blockade. Initial findings supporting the antigen-independent effects of IgE on FcεRI signaling and on the survival and proliferation of cultured mast cells were presented as back-to-back reports in 2001 (Asai et al. 2001; Kalesnikoff et al. 2001). A role for IgE antibodies in mast cell expansion in vivo is supported by our observation that the appearance of mast cells in the airways of mice subjected to repeated inhalation of *Aspergillus funigatus* was markedly impaired in animals lacking IgE antibodies (Mathias et al. 2009).

Studies of antigen-independent effects of IgE antibodies on mast cells have been performed using monoclonal IgE antibodies of varying specificity. Kawakami and colleagues observed that there is considerable variability among individual antibodies with respect to the potency of their antigen-independent effects on mast cells, and they have designated clones capable of inducing high-level antigen-independent cytokine secretion, including SPE-7, as "Highly Cytokinergic" (HC) (Kashiwakura et al. 2012). They have demonstrated that HC IgE clones exhibit reactivity to autoantigens (dsDNA, ssDNA, and thyroglobulin) an observation that provides a potential mechanism whereby cross-linking of receptor-bound IgE is induced by ubiquitous autoantigens in the absence of nominal antigen. James and colleagues published a very interesting finding in 2003 that both anticipated this autoreactivity of HC IgE antibodies and provided a structural basis (James et al. 2003). Using X-ray crystallography, they observed that the antigen-binding site of HC IgE, SPE-7, exists in equilibrium of alternative isoforms, at least one of which could bind to an autoantigen (thioredoxin) identified by peptide screening. Taken together, these observations indicate that the pro-allergic effects of IgE antibodies may be observed in the absence of allergen exposure and that IgE blockade could exert effects even in allergen-independent contexts.

1.4 Antiviral Responses

Consideration of potential impacts of IgE on antiviral responses is intriguing, as viral infections are the most common triggers of asthma. Multiple studies have demonstrated reduced production of type I interferons in atopic patients. Cells recovered from bronchoalveolar lavage, dendritic cells, and peripheral blood mononuclear cells from asthmatic patients were evaluated after treatment with viruses including influenza and rhinovirus (Gill et al. 2010; Sykes et al. 2012; Durrani et al. 2012). Dendritic cells from atopic patients produced significantly less IFN-α when exposed to influenza, and the amount produced was inversely proportional to total serum IgE (Gill et al. 2010). Supportive data were shown from BAL cells from allergic patients exposed to rhinovirus. Production of type I interferons was again deficient, and this reduced production was associated with worsening airway hyperresponsiveness (Sykes et al. 2012). In addition, cross-linking FcεRI on PBMCs prior to exposure to human rhinovirus also triggered a significant decrease in IFN-α production (Durrani et al. 2012). Such observations suggest that reductions in free IgE may help improve immune responses to viral infections and attenuate asthma exacerbations in patients exposed to respiratory viruses.

2 Omalizumab: Structure and Pharmacology

2.1 Structure: Humanized MAb Against IgE

Omalizumab (initially designated rhuMAb-E25) is a humanized IgG$_1$κ monoclonal anti-IgE antibody that binds with high affinity to IgE at the Cε3 domain and has a molecular weight of approximately 149 kD (Fig. 2). The antibody is derived from the murine monoclonal antibody, MAE11. MAE11 was humanized by incorporating the mouse-derived complementarity-determining region into a human IgG1 framework (Presta et al. 1993; Boushey 2001). The IgG1 human framework comprises approximately 95 % of the omalizumab molecule, leaving 5 % murine sequences. Importantly, omalizumab binds circulating IgE regardless of its antigen specificity. It effectively blocks the binding site of IgE for both FcεRI and CD23, and its high affinity for IgE allows it to compete with FcεRI. Preclinical studies demonstrated reduction in serum levels of free IgE in murine models (Haak-Frendscho et al. 1994), while subsequent clinical studies established this capacity to reduce IgE in asthmatic patients. Omalizumab was approved for use in the USA in 2003 and was later approved as adjunctive therapy in 2005 by the European Medicines Agency. While omalizumab remains licensed only for patients aged 12 or older in the USA, the medication was recently approved by the European Medicines Agency for use in children aged 6 years and older (Pelaia et al. 2011).

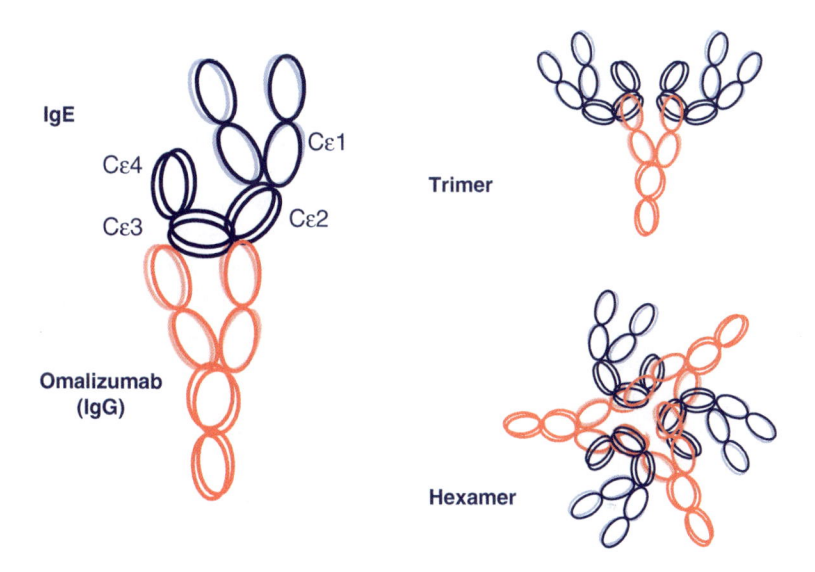

Fig. 2 Omalizumab. Omalizumab is a humanized IgG$_1$κ monoclonal anti-IgE antibody that binds to IgE at the Cε3 domain. Omalizumab forms trimeric and hexameric complexes with IgE, which circulate. IgE in these complexes is detected in standard clinical IgE assays so that total IgE levels appear to rise following institution of omalizumab therapy

2.2 IgE–Omalizumab Interaction

Omalizumab specifically binds circulating free IgE at the high-affinity receptor-binding site, thus inhibiting binding of the antibody to effector cells. Once bound to free IgE, small complexes are formed. Typically, omalizumab:IgE complexes are trimers of approximately 490–530 kD or hexamers of approximately 1,000 kD. The size of complexes depends on the relative concentrations of both serum-free IgE and omalizumab. A study of anti-IgE in cynomolgus monkeys found the largest complex formed in vivo was a hexamer (Liu et al. 1995; Fox et al. 1996). Omalizumab does not cause cross-linking of the IgE receptor or subsequent effector cell degranulation, a property that has been referred to as its being "nonanaphylactogenic" (Chang 2000).

2.3 Pharmacokinetics

Omalizumab is administered subcutaneously, with a bioavailability of 62 %. Peak serum concentrations are reached after an average of 7–8 days, and the half-life is approximately 26 days. This is in contrast to the much shorter 2-day half-life of free IgE. Omalizumab dosing utilizes important data gleaned from a phase II clinical trial of anti-IgE for the treatment of allergic rhinitis triggered by ragweed allergen (Pelaia et al. 2011). The study concluded that significant symptomatic benefit was reached only when free IgE levels were completely suppressed and that the magnitude of suppression was associated with both pretreatment serum IgE level and anti-IgE dosing. Therefore, omalizumab is administered every 2–4 weeks at a dose of 150–375 mg subcutaneously, with the dose and frequency determined based on patient body weight and IgE level. Due to the requirement of near-total suppression of IgE levels for clinical efficacy, the drug is only FDA approved for pretreatment IgE levels of 30–700 IU/mL (XOLAIR PACKAGE INSERT).

Following administration of omalizumab, serum-free IgE concentrations decrease rapidly in clinical studies, with a mean drop of over 96 % from baseline using standard dosing. However, total serum IgE including both bound and unbound fractions increased subsequent to omalizumab dosing. The clearance of omalizumab:IgE complexes is slower than that of free IgE, and standard clinical lab assays detect IgE in both free and complexed forms. Thus, an apparently paradoxical *increase* in total IgE is observed in omalizumab-treated patients. After discontinuation of omalizumab dosing, serum total IgE decreases, while free IgE increases. However, this reversal occurs slowly and does not reach pretreatment levels even one year after discontinuation of therapy.

3 Early Clinical Studies in Asthma

3.1 Inhalation Challenge

Early clinical studies demonstrated that anti-IgE treatment in asthmatic patients led to significantly decreased free IgE levels and reduced both immediate symptoms and late-phase responses to inhaled aeroallergen challenge (Fahy et al. 1997; Boulet et al. 1997).

3.2 Early and Late-phase Suppression, Further Analyses

Attenuation of the early allergen response in the above studies was demonstrated both by a decrease in the degree of the fall of FEV_1 after exposure to fixed doses of allergen and by an increase in the threshold dose of allergen necessary to elicit bronchoconstriction. Fahy et al. (1997) showed that late-phase responses in the airway were also reduced in the treatment groups, as evidenced by protection from late drops in FEV_1 and by a reduction in the percentage of airway eosinophils in sputum 24 h after challenge. These early studies set the stage for future evaluations of the safety and efficacy of anti-IgE in asthma and other atopic diseases. The demonstration that anti-IgE could counteract both early and late-phase allergic responses confirmed the importance of IgE antibody function both in immediate hypersensitivity and in cellular recruitment and the initiation of inflammation, providing a strong rationale for clinical trials of IgE blockade in chronic allergic diseases such as asthma.

4 Asthma Studies

Subsequent studies evaluating the efficacy of anti-IgE therapy for allergic asthma in adolescent and adult patients further underscored the ability of anti-IgE to attenuate early and late-phase asthma symptoms. Milgrom and colleagues completed a large randomized, placebo-controlled phase II study evaluating patients with moderate to severe allergic asthma (Milgrom et al. 1999). In treatment groups, serum-free IgE levels dropped precipitously, and asthma symptoms were significantly improved as evidenced by reduced use of rescue β-agonist inhalers, decreased oral and inhaled corticosteroid (ICS) requirement, and improved pulmonary function testing.

Two critical phase III clinical trials of anti-IgE therapy in patients with severe chronic asthma, performed in the USA and Europe, provided the evidence that led to FDA approval of omalizumab (Soler et al. 2001; Busse et al. 2001). Both groups studied ICS-dependent asthmatics and utilized a study design that included initial continuation of daily ICS, with a subsequent steroid reduction phase during anti-IgE treatment. In both trials, treatment group patients suffered fewer asthma exacerbations during both the steroid stable and reduction phases. Other primary end points in the treatment groups included a significant reduction in ICS dose, decreased frequency of rescue inhaler use, improved pulmonary function parameters including mean FEV_1, and improved asthma symptoms scores. The frequency of adverse events was not increased in the anti-IgE treatment groups.

A veritable flood of trials followed on the heels of these initial studies. The results of twenty-five of these anti-IgE studies comprising a total of 6,382 participants with mild to severe allergic asthma were analyzed in a recent Cochrane review (Normansell et al. 2014). Most of the included studies evaluated anti-IgE as adjunctive therapy to inhaled corticosteroids and/or long acting beta agonists (LABAs). Overall, the pooled data indicated that clinical benefit from anti-IgE therapy was improved from placebo, including reduced asthma exacerbation frequency, reduced hospitalizations, and reduction in daily ICS and rescue inhaler use. Also, anti-IgE therapy significantly increased the number of participants who were able to completely withdraw from daily ICS use as compared to placebo. Participants in 7 of 11 studies also noted modest but significant improvements in asthma scores and quality of life assessments. However, changes in pulmonary function parameters were inconsistent. Overall, the message of both the initial pivotal studies and the large number of follow-up trials has been that anti-IgE reduces the frequency of acute flares and hospitalizations while only modestly affecting the chronic day-to-day indicators of disease activity such as quality of life scores and physiologic measures of small airway obstruction. These findings suggest that anti-IgE attenuates disease flares precipitated by aeroallergen exposure or respiratory infection but that it does not interfere with the chronic Th2-driven allergic inflammatory process that leads to the infiltration of the airways with inflammatory cells, edema of the airway mucosa, hyper-responsiveness of bronchial smooth muscle, and hyper-production of mucus all of which can likely arise in the absence of IgE signals.

5 Pediatric Asthma

Most of the initial anti-IgE studies focused on adult participants, but in recent years more emphasis has been placed on childhood asthma, a disease that appears to be more consistently allergen driven. Milgrom and colleagues performed a large, double-blind, randomized, placebo-controlled trial to evaluate the safety and efficacy of anti-IgE in the treatment of children aged 6–12 years with moderate to severe asthma (Milgrom et al. 2001). This study design echoed adult studies, with the 334 enrolled children treated with anti-IgE as adjunctive therapy to stable ICS regimens. Inclusion criteria for pediatric participants included moderate to persistent asthma well controlled on ICS, positive skin prick testing to at least one common aeroallergen, and total IgE levels between 30 and 1,300 IU/mL. These criteria differ from the previously discussed adolescent and adult studies, where the majority of trials included patients with poorly controlled asthma despite ICS use and a more restricted pretreatment IgE level. Results were encouraging in the treatment group, with significant reductions in ICS dosage and more frequent complete withdrawal of ICS use. During the steroid reduction phase in the treatment group, use of rescue albuterol was reduced, and asthma exacerbations requiring additional treatment declined in both frequency and severity. No significant change between the groups was seen in asthma symptom scores or pulmonary function parameters. Anti-IgE was well tolerated, with no serious adverse effects. Following completion of the above trial, further information regarding quality of life evaluations was published (Lemanske et al. 2002). No change in quality of life was found during the steroid-stable phase of the trial, but at the end of the study treatment group participants reported improved asthma scores and a statistically significant increase in overall asthma-related quality of life. This was likely affected by the reduction in exacerbations, decrease in ICS use, and lack of hospitalizations in the treatment group.

A more recent pediatric anti-IgE trial performed by Lanier et al. (2009) had similar results. This study more closely resembled the adult trials, as enrolled pediatric subjects included children aged 6 to <12 years old with moderate to severe persistent asthma incompletely controlled with ICSs. Over the course of the trial, the overall exacerbation frequency was reduced by 43 % versus the placebo group. Additionally, in contrast to the previous pediatric study, treatment group subjects experienced a significant attenuation in exacerbation frequency during the fixed-steroid phase. Further support of the effectiveness of omalizumab in adjunctive treatment of pediatric asthma was exhibited by a recent retrospective analysis of pooled data extracted from five adolescent and adult anti-IgE trials (Massanari et al. 2009). This analysis included patients aged 12–17 years old and also demonstrated a reduction in frequency of asthma exacerbations, improved pulmonary function parameters, and improvement in asthma scores. In all studies, regardless of participant age, anti-IgE therapy significantly reduced asthma exacerbations, thus highlighting the importance of IgE during times of asthma exacerbation. However, as was evident in the meta-analyses of the adult trials, improvements in pulmonary function parameters and quality of life were inconsistent across studies. This again reveals the key role of IgE in acute flares while suggesting a dominant role for IgE-independent mechanisms in the pathogenesis of the chronic inflammation and impairments in airway physiology seen in asthma.

6 Omalizumab in Other Atopic Diseases

Alongside the successes of omalizumab in asthma trials, anti-IgE was evaluated early on for its effectiveness in allergic rhinitis (AR), another disease characterized by high-titer IgE responses to inhaled aeroallergens and by seasonal flares of symptoms driven by allergen inhalation (Casale et al. 1997). Subsequently, other disorders became candidates for anti-IgE trials, including atopic dermatitis, urticaria, and food allergy (Chang et al. 2007; Incorvaia et al. 2014).

6.1 Allergic Rhinitis

Allergic rhinitis is a prevalent problem in both adult and pediatric patient populations. A number of trials have evaluated use of anti-IgE in patients with allergic rhinitis triggered by a variety of environmental allergens, most commonly birch pollen or ragweed. Their results established that anti-IgE decreased nasal and ocular symptoms, reduced antihistamine use, and improved quality of life scores (Adelroth et al. 2000; Casale et al. 1997, 2001; Chervinsky et al. 2003). Anti-IgE also improved symptom scores and decreased sensitivity to nasal allergen challenge (Hanf et al. 2004). Nasal biopsy specimens in patients receiving anti-IgE showed none of the mucosal tissue eosinophilia evident in the placebo group (Plewako et al. 2002). Further evidence for improvement in allergic rhinitis was demonstrated by a trial of anti-IgE administered in combination with traditional subcutaneous immunotherapy in children and adolescents with moderate to severe allergic rhinitis. Study participants receiving anti-IgE in addition to standard allergen immunotherapy showed significant clinical improvement in symptoms as compared to immunotherapy alone (Kuehr et al. 2002). Thus, removal of IgE seems to ameliorate both early and late-phase hypersensitivity responses in allergic rhinitis.

6.2 Rush Immunotherapy

Rush immunotherapy is an accelerated subcutaneous immunotherapy regimen that allows patients to reach maintenance dosing much more quickly than with traditional protocols. However, these faster protocols are associated with an increased incidence of serious adverse reactions including anaphylaxis. Casale et al. (2006) established that adult patients treated with anti-IgE prior to induction of rush immunotherapy had fewer reactions. Another study showed addition of anti-IgE prior to cluster immunotherapy in patients with persistent asthma reduced the risk of adverse events during immunotherapy. More patients in the anti-IgE treatment group reached goal maintenance dosing and had fewer adverse events, including those requiring epinephrine, throughout the cluster protocol (Massanari et al. 2010).

6.3 Food-induced Anaphylaxis and Oral Immunotherapy

Food allergy has emerged as another attractive option for anti-IgE therapy. Immediate hypersensitivity reactions to allergenic food ingestion depend on the presence of food-specific IgE antibodies and can be quite severe, progressing to potentially fatal systemic anaphylaxis. As current therapeutic options are limited to strict avoidance of the allergenic food for safety reasons, patients are paradoxically deprived of their only hope for achieving tolerance, namely oral ingestion of the allergenic food. Anti-IgE has been considered an attractive treatment option in this setting both because it might reduce dangerous reactions following inadvertent food ingestions and as an adjunct to oral immunotherapy.

The first study evaluating use of anti-IgE in food allergy was conducted by Leung et al. (2003) who showed that administration of an anti-IgE monoclonal antibody, TNX-901, led to significant increases in the threshold dose of peanut required to trigger hypersensitivity reactions in oral peanut food challenges. This result established the efficacy of anti-IgE in blunting IgE-mediated food reactions and provided the basis for the more recent application of anti-IgE treatment during oral immunotherapy (OIT) in children with food allergies. While OIT, in which allergic subjects are fed incrementally increasing doses of allergenic food under very controlled conditions, has shown promise in terms of reducing food sensitivity, its broad application has been impeded by the frequency of undesirable and potentially dangerous reactions experienced by patients in the course of dose escalation. Nadeau et al. reasoned that anti-IgE might mitigate these reactions and conducted a pilot study in children with cow's milk allergy undergoing rapid oral milk desensitization under cover of anti-IgE. Nine of 10 patients reached the goal dose of 2,000 mg milk after desensitization over 7–11 weeks. Anti-IgE therapy was then discontinued, and daily milk ingestion continued. All 9 patients subsequently passed a double-blind, placebo-controlled food challenge (DBPCFC) 8 weeks after discontinuation of anti-IgE (Nadeau et al. 2011).

Anti-IgE therapy has also been applied to OIT for the extremely allergenic food, peanut. A phase II trial performed by Sampson et al. (2011) to evaluate the effectiveness of anti-IgE in reducing the frequency of peanut-induced allergic reactions during OIT was discontinued early due to anaphylactic episodes that occurred during initial pre-enrollment peanut food challenges. However, data obtained prior to discontinuation revealed that participants receiving anti-IgE therapy were able to tolerate higher peanut doses than placebo-treated participants. Another study evaluating peanut oral immunotherapy under cover of anti-IgE has been performed (Schneider et al. 2013). High-risk peanut allergic children were treated with anti-IgE, and then underwent oral desensitization. Patients reached a cumulative dose of 992 mg without symptoms of hypersensitivity, and subsequently tolerated a peanut food challenge of 8,000 mg peanut flour. These studies illuminate the possibilities of future successful, safe oral desensitization to foods. It will be of great interest to determine whether IgE blockade during OIT, in addition

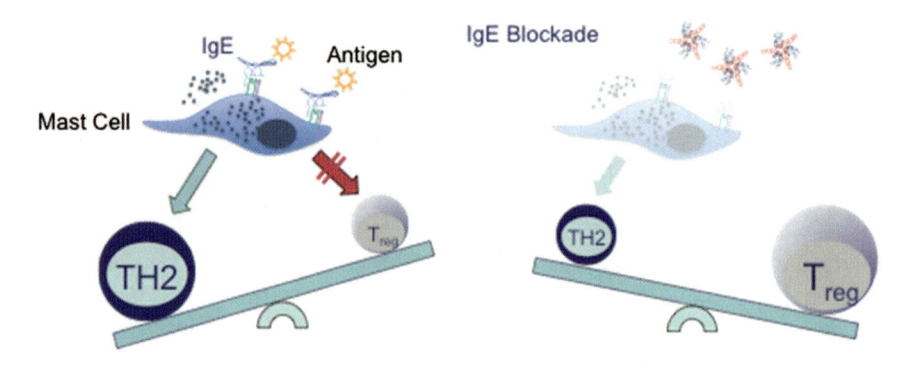

Fig. 3 Hypothetical model for IgE-mediated regulation of T cell responses to allergen and effects of omalizumab in this model. Mast cells produce significant levels of the Th2-skewing cytokine IL-4. We speculate that IgE-mediated mast cell activation drives IL-4-mediated Th2 expansion and concomitant Treg suppression and that inhibition of this effect in the setting of omalizumab therapy would favor Treg responses over Th2

to enhancing safety by reducing anaphylactic reactions, might also modulate immune responses to food allergens, suppressing allergen-driven Th2 expansion, and favoring Treg induction (Fig. 3).

6.4 Chronic Idiopathic Urticaria

Chronic idiopathic urticaria (CIU) is a condition characterized by hives that last for a minimum of 6 weeks, without obvious trigger. Therapeutic options currently include antihistamines, leukotriene receptor antagonists, systemic steroids, and other immunosuppressive medications. However, the condition is often refractory to these therapies and the side effects of systemic steroids are undesirable. An early proof of concept study using anti-IgE demonstrated improvement in the Urticaria Activity Score (UAS), which includes pruritus severity, number of hives, and size of largest hive, as well as a reduction in rescue antihistamine use and improvement in overall quality of life (Kaplan et al. 2008). Subsequent phase II and III trials evaluating the efficacy of anti-IgE compared to placebo in treating CIU unresponsive to traditional therapies substantiated the findings of previous proof of concept trials. The first phase III trial performed by Maurer et al. (2013) included 323 patients with treatment-resistant moderate to severe CIU and evaluated efficacy of increasing doses of omalizumab.

Since CIU is not generally considered an atopic disease and allergen triggers are not usually identified in these patients, the efficacy of omalizumab was a surprise to some. The average IgE level of the study population was only 168.2 ± 231.9 IU/ml, which is only very modestly elevated and much lower than the levels observed in the other atopic conditions responsive to anti-IgE. The study found significant dose responsive diminution of symptoms as compared to placebo, with the greatest clinical effects generated by the highest anti-IgE dose (300 mg). A subsequent phase III trial performed by Kaplan et al. (2013) substantiated these results. Omalizumab was recently approved by the FDA for the treatment of CIU refractory to antihistamine therapy in patients aged 12 years and older.

The mechanisms by which anti-IgE reduces symptoms in patients with CIU remain to be elucidated, but a rapid reduction in free IgE levels has been consistently demonstrated in omalizumab-treated patients with CIU and this reduction in IgE correlates closely with clinical improvement. Several mechanistic studies of omalizumab have demonstrated downregulation of FcεRI on peripheral blood basophils and on skin and nasal mast cells (Saini et al. 1999; Beck et al. 2004; Eckman et al. 2010). There is some evidence that a subset of patients with CIU has autoantibodies directed against FcεRI (Grattan et al. 1991; Hide et al. 1993; Kikuchi and Kaplan 2001), and it has been speculated that the reduction in FcεRI density induced by omalizumab therapy reduces the susceptibility of FcεRI$^+$ cells to activation by such antibodies. An alternative possibility is that these patients have HC IgE antibodies (see Sect. 1.3) that interact with autoantigens and omalizumab decreases the titers of these antibodies. While both the results of the CIU trials and these proposed mechanisms are thought provoking, further studies are required before firm conclusions can be made.

6.5 Atopic Dermatitis

Atopic dermatitis (AD) typically presents in childhood and can persist throughout the lifetime of an affected patient. Patients typically have dramatically elevated serum IgE levels and are often affected by other concurrent allergic diseases suggesting a potential pathogenic role of IgE in AD. Treatment of recalcitrant AD can be difficult and may include immunosuppressive medications. Clinical studies of anti-IgE therapy for AD have resulted in variable conclusions. Case reports and an early pilot study using omalizumab suggested improvement in clinical atopic dermatitis symptoms (Sheinkopf et al. 2008). A larger explorative study also found treatment with anti-IgE significantly reduced free serum IgE, decreased FcεRI expression on basophil cell surfaces, and also decreased IgE saturation of FcεRI. However, treatment group patients had no improvement in clinical symptoms (Heil et al. 2010). These findings, like those of the asthma studies, suggest that the chronic inflammatory component of AD is likely driven in large part by IgE-independent mechanisms and that the utility of IgE blockade may be limited in this setting.

6.6 Eosinophilic Gastrointestinal Disorders

Eosinophilic gastrointestinal disorders include eosinophilic esophagitis, eosinophilic colitis, and eosinophilic gastroenteritis. Affected patients experience a range of chronic gastrointestinal symptoms including difficulty swallowing, pain, and diarrhea but, unlike typical food allergy patients, they do not generally experience immediate hypersensitivity reactions, such as systemic anaphylaxis. In this way, the disease process seems more analogous to that involved in AD. However, eosinophilic gastrointestinal disorders can be associated with atopy, and some patients have positive skin prick testing to foods. A clinical trial evaluating the effects of anti-IgE therapy in patients with eosinophilic gastroenteritis gave results consistent with those for anti-IgE therapy for other atopic conditions (Foroughi et al. 2007). As expected, serum-free IgE and food allergen skin prick test reactions were decreased in the treatment group along with reductions in FcεRI surface expression on basophils and dendritic cells. A downward trend in tissue eosinophilia was observed in biopsies of the duodenum, gastric antrum, and gastric body but this did not reach significance. Symptom scores improved both midstudy and at the conclusion of the trial, but did not correlate with reduction in tissue eosinophilia.

7 Mechanisms

Many of the clinical trials of IgE blockade described above have been accompanied by immunologic analyses, which have provided valuable insights both into the pathogenesis of allergic diseases and the mechanisms of action of anti-IgE.

7.1 Blockade of Early and Late-phase Hypersensitivity

The seminal studies of anti-IgE therapy have demonstrated sharp reductions in circulating levels of free IgE antibodies along with strong evidence that anti-IgE blockade impairs both early and late-phase hypersensitivity reactions. Anti-IgE therapy reduces early and late-phase allergic asthma responses after inhalational allergen challenge (Fahy et al. 1997), diminishes late-phase skin responses to intradermal allergen exposure (Ong et al. 2005), and decreases sputum eosinophilia (Fahy et al. 1997; Djukanovic et al. 2004) and airway submucosal eosinophil, T cell, and B cell numbers in asthmatic subjects (Djukanovic et al. 2004). A study performed on adult subjects with allergic rhinitis secondary to dust mites showed a rapid reduction in serum-free IgE levels and a 97 % decrease of basophil cell surface expression of FcεRI (MacGlashan et al. 1997). Basophil histamine release was also decreased by approximately 90 %, and basophil response was completely eradicated in 50 % of subjects. Mast cell involvement was implicated by requirement of antigen doses of 100 times the original dose in order to generate similar pre-anti-IgE skin prick wheals. Peripheral eosinophilia and serum levels of Th2-inducing cytokines including IL-13 are also decreased in asthmatic patients treated with anti-IgE (Noga et al. 2003). The latter observation is consistent with a role for IgE antibodies in the maintenance of Th2-driven allergic responses and highlights the importance of anti-IgE therapy in combating these responses (Fig. 3).

7.2 Regulation of IgE Receptors

Prior to the application of anti-IgE therapy, it was known that IgE levels regulate expression and stability of FcεRI receptors on the surface of basophils, mast cells, and dendritic cells. Treatment with anti-IgE depletes the concentration of free IgE in serum, and therefore, as predicted, results in decreased cell surface density of FcεRI (MacGlashan et al. 1997; Beck et al. 2004; Prussin et al. 2003). Beck and colleagues evaluated the effects of omalizumab treatment on mast cell FcεRI expression. After 1 week of therapy, minimal reduction in mast cell surface FcεRI was noted, and later in therapy FcεRI receptor density was significantly decreased, along with a reduction in skin prick allergen-stimulated wheal size (Beck et al. 2004). While these studies have been interpreted to imply discordance in rate of FcεRI reduction between mast cells and basophils, there are technical differences in the approach to FcεRI measurement that limit this extrapolation. Basophil FcεRI density has typically been quantitated by flow cytometry, a method that reliably detects only cell surface receptor while the measurements of FcεRI on skin mast cells have been performed using immunohistochemical staining which does not discriminate surface from internalized FcεRI. In analyses of cultured rodent mast cells, surface FcεRI levels modulate very rapidly in response to changes in ambient IgE. It will be important in future studies to determine the kinetics of reduction of surface levels of FcεRI on mast cells of omalizumab-treated subjects.

Dendritic cells, which also express FcεRI, are key inducers and regulators of immune responses. Omalizumab therapy results in reduction of FcεRI expression on circulating plasmacytoid dendritic cells (pDCs) as demonstrated by Prussin et al. (2003). The reduction of pDC surface FcεRI receptor density was evident within 7 days of anti-IgE therapy and correlated well with reduced serum-free IgE levels. Schroeder et al. (2010) have reported that the reduction in DC FcεRI following omalizumab therapy is associated with a reduction in IgE-facilitated antigen presentation in the induction of Th2 responses in a cell culture system. These results are of great interest and suggest that anti-IgE therapy may interfere with IgE-facilitated antigen presentation by dendritic cells. This ultimately could cause a blockade of early sensitization to antigens. This in turn acts in concert with reduction of effector cell mediation of immediate hypersensitivity responses to further reduce hypersensitivity reactions.

7.3 Effects on Viral Immunity

Upper respiratory viral infections are probably the most important trigger of asthma exacerbations across age groups. While anti-IgE therapy has proven to be effective in decreasing the frequency of asthma exacerbations overall, it has been unclear whether this is exclusively because of interference with immediate hypersensitivity reactions in patients exposed to inhaled aeroallergens or whether omalizumab might also alter the onset of asthma flares after viral infection. In the "Inner City Asthma Study," (Busse et al. 2011) made the intriguing observation that anti-IgE treatment is associated a significant reduction of the frequency of asthma flares associated with seasonal virus exposure. In a substudy of this cohort, the investigators found that rates of viral infection as measured in nasal swabs were the same in treatment and placebo groups, suggesting a benefit of anti-IgE treatment in attenuating virus-induced asthma flares. We speculate that this resistance to flares might be due either to an elevated threshold for bronchial obstruction because of reduced allergic inflammation of the airway mucosa in omalizumab-treated subjects or to interference with the negative effects of IgE:FcεRI signaling on innate immune responses (Type I interferon production) to viruses by dendritic cells (see Sect. 1.4).

8 Future Directions

The success of omalizumab in the treatment of allergic asthma has generated new insights into the roles of IgE in allergic pathophysiology and has opened doors for novel strategies. The generation of higher affinity anti-IgE reagents is in progress (Cohen et al. 2014) as is the design of therapeutics, which uniquely bind to IgE$^+$ B cells and could selectively target IgE production at the source (Brightbill et al. 2010). There is considerable interest in the development of inhibitors of the very high-affinity interaction of IgE with FcεRI. Jardetzky and colleagues have presented encouraging data regarding designed ankyrin repeat inhibitors (DARPins), which have the potential to dislodge IgE bound to FcεRI (Eggel et al. 2014). Other promising studies have focused on blocking critical activators of key IgE-mediated signaling pathways. Spleen Tyrosine Kinase (SYK) is the critical proximal protein tyrosine kinase in FcεRI signaling. Early studies have shown inhibition of SYK leads to mitigation of airway inflammation and mast cell degranulation (Matsubara et al. 2006; Penton et al. 2013; Yamamoto et al. 2003). Moy et al. utilized a small molecule and very highly selective inhibitor of SYK (SYKi) to achieve blockade of allergen-induced bronchoconstriction in murine models (Moy et al. 2013). Meanwhile, ongoing analyses of omalizumab efficacy and mechanisms of action in a range of allergic diseases promise to further illuminate the role of IgE in sensitization and pathogenesis of allergy.

References

Acharya M, Borland G, Edkins AL, Maclellan LM, Matheson J, Ozanne BW, Cushley W (2010) CD23/Fc epsilon RII: molecular multi-tasking. Clin Exp Immunol 162(1):12–23. doi:10.1111/j.1365-2249.2010.04210.x

Adelroth E, Rak S, Haahtela T, Aasand G, Rosenhall L, Zetterstrom O, Byrne A, Champain K, Thirlwell J, Cioppa GD, Sandstrom T (2000) Recombinant humanized mAb-E25, an anti-IgE mAb, in birch pollen-induced seasonal allergic rhinitis. J Allergy Clin Immunol 106(2):253–259. doi:10.1067/mai.2000.108310

Asai K, Kitaura J, Kawakami Y, Yamagata N, Tsai M, Carbone DP, Liu FT, Galli SJ, Kawakami T (2001) Regulation of mast cell survival by IgE. Immunity 14(6):791–800

Beck LA, Marcotte GV, MacGlashan D, Togias A, Saini S (2004) Omalizumab-induced reductions in mast cell Fc epsilon RI expression and function. J Allergy Clin Immunol 114(3):527–530

Borkowski TA, Jouvin MH, Lin SY, Kinet JP (2001) Minimal requirements for IgE-mediated regulation of surface Fc epsilon RI. J Immunol 167(3):1290–1296

Boulet LP, Chapman KR, Cote J, Kalra S, Bhagat R, Swystun VA, Laviolette M, Cleland LD, Deschesnes F, Su JQ, DeVault A, Fick RB Jr, Cockcroft DW (1997) Inhibitory effects of an anti-IgE antibody E25 on allergen-induced early asthmatic response. Am J Respir Crit Care Med 155(6):1835–1840. doi:10.1164/ajrccm.155.6.9196083

Boushey HA Jr (2001) Experiences with monoclonal antibody therapy for allergic asthma. J Allergy Clin Immunol 108(2 Suppl):S77–83

Brightbill HD, Jeet S, Lin Z, Yan D, Zhou M, Tan M, Nguyen A, Yeh S, Delarosa D, Leong SR, Wong T, Chen Y, Ultsch M, Luis E, Ramani SR, Jackman J, Gonzalez L, Dennis MS, Chuntharapai A, DeForge L, Meng YG, Xu M, Eigenbrot C, Lee WP, Refino CJ, Balazs M, Wu LC (2010) Antibodies specific for a segment of human membrane IgE deplete IgE-producing B cells in humanized mice. J Clin Invest 120(6):2218–2229. doi:10.1172/JCI40141

Busse W, Corren J, Lanier BQ, McAlary M, Fowler-Taylor A, Cioppa GD, van As A, Gupta N (2001) Omalizumab, anti-IgE recombinant humanized monoclonal antibody, for the treatment of severe allergic asthma. J Allergy Clin Immunol 108(2):184–190

Busse WW, Morgan WJ, Gergen PJ, Mitchell HE, Gern JE, Liu AH, Gruchalla RS, Kattan M, Teach SJ, Pongracic JA, Chmiel JF, Steinbach SF, Calatroni A, Togias A, Thompson KM, Szefler SJ, Sorkness CA (2011) Randomized trial of omalizumab (anti-IgE) for asthma in inner-city children. N Engl J Med 364(11):1005–1015. doi:10.1056/NEJMoa1009705

Cameron L, Gounni AS, Frenkiel S, Lavigne F, Vercelli D, Hamid Q (2003) S epsilon S mu and S epsilon S gamma switch circles in human nasal mucosa following ex vivo allergen challenge: evidence for direct as well as sequential class switch recombination. J Immunol 171(7):3816–3822

Casale TB, Bernstein IL, Busse WW, LaForce CF, Tinkelman DG, Stoltz RR, Dockhorn RJ, Reimann J, Su JQ, Fick RB Jr, Adelman DC (1997) Use of an anti-IgE humanized monoclonal antibody in ragweed-induced allergic rhinitis. J Allergy Clin Immunol 100(1):110–121

Casale TB, Condemi J, LaForce C, Nayak A, Rowe M, Watrous M, McAlary M, Fowler-Taylor A, Racine A, Gupta N, Fick R, Della Cioppa G (2001) Effect of omalizumab on symptoms of seasonal allergic rhinitis: a randomized controlled trial. JAMA 286(23):2956–2967

Casale TB, Busse WW, Kline JN, Ballas ZK, Moss MH, Townley RG, Mokhtarani M, Seyfert-Margolis V, Asare A, Bateman K, Deniz Y (2006) Omalizumab pretreatment decreases acute reactions after rush immunotherapy for ragweed-induced seasonal allergic rhinitis. J Allergy Clin Immunol 117(1):134–140. doi:10.1016/j.jaci.2005.09.036

Chang TW (2000) The pharmacological basis of anti-IgE therapy. Nat Biotechnol 18(2):157–162

Chang TW, Wu PC, Hsu CL, Hung AF (2007) Anti-IgE antibodies for the treatment of IgE-mediated allergic diseases. Adv Immunol 93:63–119. doi:10.1016/S0065-2776(06)93002-8

Chervinsky P, Casale T, Townley R, Tripathy I, Hedgecock S, Fowler-Taylor A, Shen H, Fox H (2003) Omalizumab, an anti-IgE antibody, in the treatment of adults and adolescents with perennial allergic rhinitis. Ann Allergy Asthma Immunol 91(2):160–167. doi:10.1016/S1081-1206(10)62171-0

Cohen ES, Dobson CL, Kack H, Wang B, Sims DA, Lloyd CO, England E, Rees DG, Guo H, Karagiannis SN, O'Brien S, Persdotter S, Ekdahl H, Butler R, Keyes F, Oakley S, Carlsson M, Briend E, Wilkinson T, Anderson IK, Monk PD, von Wachenfeldt K, Eriksson PO, Gould HJ, Vaughan TJ, May RD (2014) A novel IgE-neutralizing antibody for the treatment of severe uncontrolled asthma. MAbs 6(3):1

Djukanovic R, Wilson SJ, Kraft M, Jarjour NN, Steel M, Chung KF, Bao W, Fowler-Taylor A, Matthews J, Busse WW, Holgate ST, Fahy JV (2004) Effects of treatment with anti-immunoglobulin E antibody omalizumab on airway inflammation in allergic asthma. Am J Respir Crit Care Med 170(6):583–593. doi:10.1164/rccm.200312-1651OC

Dombrowicz D, Quatannens B, Papin JP, Capron A, Capron M (2000) Expression of a functional Fc epsilon RI on rat eosinophils and macrophages. J Immunol 165(3):1266–1271

Durrani SR, Montville DJ, Pratt AS, Sahu S, DeVries MK, Rajamanickam V, Gangnon RE, Gill MA, Gern JE, Lemanske RF Jr, Jackson DJ (2012) Innate immune responses to rhinovirus are reduced by the high-affinity IgE receptor in allergic asthmatic children. J Allergy Clin Immunol 130(2):489–495. doi:10.1016/j.jaci.2012.05.023

Eckman JA, Sterba PM, Kelly D, Alexander V, Liu MC, Bochner BS, Macglashan DW Jr, Saini SS (2010) Effects of omalizumab on basophil and mast cell responses using an intranasal cat allergen challenge. J Allergy Clin Immunol 125(4):889–895. doi:10.1016/j.jaci.2009.09.012 e887

Eggel A, Baravalle G, Hobi G, Kim B, Buschor P, Forrer P, Shin JS, Vogel M, Stadler BM, Dahinden CA, Jardetzky TS (2014) Accelerated dissociation of IgE-Fc epsilon RI complexes by disruptive inhibitors actively desensitizes allergic effector cells. J Allergy Clin Immunol 133(6):1709–1719. doi:10.1016/j.jaci.2014.02.005

Fahy JV, Fleming HE, Wong HH, Liu JT, Su JQ, Reimann J, Fick RB Jr, Boushey HA (1997) The effect of an anti-IgE monoclonal antibody on the early-and late-phase responses to allergen inhalation in asthmatic subjects. Am J Respir Crit Care Med 155(6):1828–1834. doi:10.1164/ajrccm.155.6.9196082

Foroughi S, Foster B, Kim N, Bernardino LB, Scott LM, Hamilton RG, Metcalfe DD, Mannon PJ, Prussin C (2007) Anti-IgE treatment of eosinophil-associated gastrointestinal disorders. J Allergy Clin Immunol 120(3):594–601. doi:10.1016/j.jaci.2007.06.015

Fox JA, Hotaling TE, Struble C, Ruppel J, Bates DJ, Schoenhoff MB (1996) Tissue distribution and complex formation with IgE of an anti-IgE antibody after intravenous administration in cynomolgus monkeys. J Pharmacol Exp Ther 279(2):1000–1008

Gill MA, Bajwa G, George TA, Dong CC, Dougherty II, Jiang N, Gan VN, Gruchalla RS (2010) Counterregulation between the Fc epsilon RI pathway and antiviral responses in human plasmacytoid dendritic cells. J Immunol 184(11):5999–6006. doi:10.4049/jimmunol.0901194

Gould HJ, Sutton BJ (2008) IgE in allergy and asthma today. Nat Rev Immunol 8(3):205–217. doi:10.1038/nri2273

Gould HJ, Sutton BJ, Beavil AJ, Beavil RL, McCloskey N, Coker HA, Fear D, Smurthwaite L (2003) The biology of IGE and the basis of allergic disease. Annu Rev Immunol 21:579–628

Grattan CE, Francis DM, Hide M, Greaves MW (1991) Detection of circulating histamine releasing autoantibodies with functional properties of anti-IgE in chronic urticaria. Clin Exp Allergy 21(6):695–704

Grayson MH, Cheung D, Rohlfing MM, Kitchens R, Spiegel DE, Tucker J, Battaile JT, Alevy Y, Yan L, Agapov E, Kim EY, Holtzman MJ (2007) Induction of high-affinity IgE receptor on lung dendritic cells during viral infection leads to mucous cell metaplasia. J Exp Med 204(11):2759–2769

Haak-Frendscho M, Robbins K, Lyon R, Shields R, Hooley J, Schoenhoff M, Jardieu P (1994) Administration of an anti-IgE antibody inhibits CD23 expression and IgE production in vivo. Immunol 82(2):306–313

Hanf G, Noga O, O'Connor A, Kunkel G (2004) Omalizumab inhibits allergen challenge-induced nasal response. Eur Respir J 23(3):414–418

Heil PM, Maurer D, Klein B, Hultsch T, Stingl G (2010) Omalizumab therapy in atopic dermatitis: depletion of IgE does not improve the clinical course—a randomized, placebo-controlled and double blind pilot study. J Dtsch Dermatol Ges 8(12):990–998. doi:10.1111/j.1610-0387.2010. 07497.x

Hide M, Francis DM, Grattan CE, Hakimi J, Kochan JP, Greaves MW (1993) Autoantibodies against the high-affinity IgE receptor as a cause of histamine release in chronic urticaria. N Engl J Med 328(22):1599–1604. doi:10.1056/NEJM199306033282204

Incorvaia C, Mauro M, Russello M, Formigoni C, Riario-Sforza GG, Ridolo E (2014) Omalizumab, an anti-immunoglobulin E antibody: state of the art. Drug Des Devel Ther 8:197–207. doi:10.2147/DDDT.S49409

James LC, Roversi P, Tawfik DS (2003) Antibody multispecificity mediated by conformational diversity. Science 299(5611):1362–1367. doi:10.1126/science.1079731

Kalesnikoff J, Huber M, Lam V, Damen JE, Zhang J, Siraganian RP, Krystal G (2001) Monomeric IgE stimulates signaling pathways in mast cells that lead to cytokine production and cell survival. Immunity 14(6):801–811

Kaplan AP, Joseph K, Maykut RJ, Geba GP, Zeldin RK (2008) Treatment of chronic autoimmune urticaria with omalizumab. J Allergy Clin Immunol 122(3):569–573. doi:10.1016/j.jaci.2008. 07.006

Kaplan A, Ledford D, Ashby M, Canvin J, Zazzali JL, Conner E, Veith J, Kamath N, Staubach P, Jakob T, Stirling RG, Kuna P, Berger W, Maurer M, Rosen K (2013) Omalizumab in patients with symptomatic chronic idiopathic/spontaneous urticaria despite standard combination therapy. J Allergy Clin Immunol 132(1):101–109. doi:10.1016/j.jaci.2013.05.013

Kashiwakura J, Okayama Y, Furue M, Kabashima K, Shimada S, Ra C, Siraganian RP, Kawakami Y, Kawakami T (2012) Most highly cytokinergic IgEs have polyreactivity to autoantigens. Allergy Asthma Immunol Res 4(6):332–340. doi:10.4168/aair.2012.4.6.332

Kikuchi Y, Kaplan AP (2001) Mechanisms of autoimmune activation of basophils in chronic urticaria. J Allergy Clin Immunol 107(6):1056–1062. doi:10.1067/mai.2001.115484

Kisselgof AB, Oettgen HC (1998) The expression of murine B cell CD23, in vivo, is regulated by its ligand, IgE. Int Immunol 10(9):1377–1384

Kraft S, Kinet JP (2007) New developments in Fc epsilon RI regulation, function and inhibition. Nat Rev Immunol 7(5):365–378. doi:10.1038/nri2072

Kubo S, Nakayama T, Matsuoka K, Yonekawa H, Karasuyama H (2003) Long term maintenance of IgE-mediated memory in mast cells in the absence of detectable serum IgE. J Immunol 170 (2):775–780

Kuehr J, Brauburger J, Zielen S, Schauer U, Kamin W, Von Berg A, Leupold W, Bergmann KC, Rolinck-Werninghaus C, Grave M, Hultsch T, Wahn U (2002) Efficacy of combination treatment with anti-IgE plus specific immunotherapy in polysensitized children and adolescents with seasonal allergic rhinitis. J Allergy Clin Immunol 109(2):274–280

Lanier B, Bridges T, Kulus M, Taylor AF, Berhane I, Vidaurre CF (2009) Omalizumab for the treatment of exacerbations in children with inadequately controlled allergic (IgE-mediated) asthma. J Allergy Clin Immunol 124(6):1210–1216. doi:10.1016/j.jaci.2009.09.021

Lemanske RF Jr, Nayak A, McAlary M, Everhard F, Fowler-Taylor A, Gupta N (2002) Omalizumab improves asthma-related quality of life in children with allergic asthma. Pediatrics 110(5):e55

Leung DY, Sampson HA, Yunginger JW, Burks AW Jr, Schneider LC, Wortel CH, Davis FM, Hyun JD, Shanahan WR Jr (2003) Effect of anti-IgE therapy in patients with peanut allergy. N Engl J Med 348(11):986–993

Lin S, Cicala C, Scharenberg AM, Kinet JP (1996) The Fc(epsilon)RI beta subunit functions as an amplifier of Fc(epsilon)RI gamma-mediated cell activation signals. Cell 85(7):985–995

Liu J, Lester P, Builder S, Shire SJ (1995) Characterization of complex formation by humanized anti-IgE monoclonal antibody and monoclonal human IgE. Biochemistry 34(33):10474–10482

MacGlashan D Jr (2008) IgE receptor and signal transduction in mast cells and basophils. Curr Opin Immunol 20(6):717–723. doi:10.1016/j.coi.2008.08.004

MacGlashan DW Jr, Bochner BS, Adelman DC, Jardieu PM, Togias A, McKenzie-White J, Sterbinsky SA, Hamilton RG, Lichtenstein LM (1997) Down-regulation of Fc(epsilon)RI expression on human basophils during in vivo treatment of atopic patients with anti-IgE antibody. J Immunol 158(3):1438–1445

Massanari M, Milgrom H, Pollard S, Maykut RJ, Kianifard F, Fowler-Taylor A, Geba GP, Zeldin RK (2009) Adding omalizumab to the therapy of adolescents with persistent uncontrolled moderate–severe allergic asthma. Clin Pediatr (Phila) 48(8):859–865. doi:10.1177/0009922809339054

Massanari M, Nelson H, Casale T, Busse W, Kianifard F, Geba GP, Zeldin RK (2010) Effect of pretreatment with omalizumab on the tolerability of specific immunotherapy in allergic asthma. J Allergy Clin Immunol 125(2):383–389. doi:10.1016/j.jaci.2009.11.022

Mathias CB, Freyschmidt EJ, Caplan B, Jones T, Poddighe D, Xing W, Harrison KL, Gurish MF, Oettgen HC (2009) IgE influences the number and function of mature mast cells, but not progenitor recruitment in allergic pulmonary inflammation. J Immunol 182(4):2416–2424. doi:10.4049/jimmunol.0801569 [pii] 182/4/2416

Matsubara S, Li G, Takeda K, Loader JE, Pine P, Masuda ES, Miyahara N, Miyahara S, Lucas JJ, Dakhama A, Gelfand EW (2006) Inhibition of spleen tyrosine kinase prevents mast cell activation and airway hyperresponsiveness. Am J Respir Crit Care Med 173(1):56–63. doi:10.1164/rccm.200503-361OC

Maurer M, Rosen K, Hsieh HJ, Saini S, Grattan C, Gimenez-Arnau A, Agarwal S, Doyle R, Canvin J, Kaplan A, Casale T (2013) Omalizumab for the treatment of chronic idiopathic or spontaneous urticaria. N Engl J Med 368(10):924–935. doi:10.1056/NEJMoa1215372

Milgrom H, Fick RB Jr, Su JQ, Reimann JD, Bush RK, Watrous ML, Metzger WJ (1999) Treatment of allergic asthma with monoclonal anti-IgE antibody. rhuMAb-E25 study group. N Engl J Med 341(26):1966–1973

Milgrom H, Berger W, Nayak A, Gupta N, Pollard S, McAlary M, Taylor AF, Rohane P (2001) Treatment of childhood asthma with anti-immunoglobulin E antibody (omalizumab). Pediatrics 108(2):E36

Moy LY, Jia Y, Caniga M, Lieber G, Gil M, Fernandez X, Sirkowski E, Miller R, Alexander JP, Lee HH, Shin JD, Ellis JM, Chen H, Wilhelm A, Yu H, Vincent S, Chapman RW, Kelly N, Hickey E, Abraham WM, Northrup A, Miller T, Houshyar H, Crackower MA (2013) Inhibition of spleen tyrosine kinase attenuates allergen-mediated airway constriction. Am J Respir Cell Mol Biol 49(6):1085–1092. doi:10.1165/rcmb.2013-0200OC

Nadeau KC, Schneider LC, Hoyte L, Borras I, Umetsu DT (2011) Rapid oral desensitization in combination with omalizumab therapy in patients with cow's milk allergy. J Allergy Clin Immunol 127(6):1622–1624. doi:10.1016/j.jaci.2011.04.009

Noga O, Hanf G, Kunkel G (2003) Immunological and clinical changes in allergic asthmatics following treatment with omalizumab. Int Arch Allergy Immunol 131(1):46–52 doi: 70434

Normansell R, Walker S, Milan SJ, Walters EH, Nair P (2014) Omalizumab for asthma in adults and children. Cochrane Database Syst Rev 1:CD003559. doi:10.1002/14651858.CD003559.pub4

Ong YE, Menzies-Gow A, Barkans J, Benyahia F, Ou TT, Ying S, Kay AB (2005) Anti-IgE (omalizumab) inhibits late-phase reactions and inflammatory cells after repeat skin allergen challenge. J Allergy Clin Immunol 116(3):558–564. doi:10.1016/j.jaci.2005.05.035

Pelaia G, Gallelli L, Renda T, Romeo P, Busceti MT, Grembiale RD, Maselli R, Marsico SA, Vatrella A (2011) Update on optimal use of omalizumab in management of asthma. J Asthma Allergy 4:49–59. doi:10.2147/JAA.S14520

Penton PC, Wang X, Amatullah H, Cooper J, Godri K, North ML, Khanna N, Scott JA, Chow CW (2013) Spleen tyrosine kinase inhibition attenuates airway hyperresponsiveness and pollution-induced enhanced airway response in a chronic mouse model of asthma. J Allergy Clin Immunol 131(2):512–520. doi:10.1016/j.jaci.2012.07.039 e511-510

Plewako H, Arvidsson M, Petruson K, Oancea I, Holmberg K, Adelroth E, Gustafsson H, Sandstrom T, Rak S (2002) The effect of omalizumab on nasal allergic inflammation. J Allergy Clin Immunol 110(1):68–71

Presta LG, Lahr SJ, Shields RL, Porter JP, Gorman CM, Fendly BM, Jardieu PM (1993) Humanization of an antibody directed against IgE. J Immunol 151(5):2623–2632

Prussin C, Griffith DT, Boesel KM, Lin H, Foster B, Casale TB (2003) Omalizumab treatment downregulates dendritic cell Fc epsilon RI expression. J Allergy Clin Immunol 112(6):1147–1154

Saini SS, MacGlashan DW Jr, Sterbinsky SA, Togias A, Adelman DC, Lichtenstein LM, Bochner BS (1999) Down-regulation of human basophil IgE and FC epsilon RI alpha surface densities and mediator release by anti-IgE-infusions is reversible in vitro and in vivo. J Immunol 162 (9):5624–5630

Saini SS, Klion AD, Holland SM, Hamilton RG, Bochner BS, Macglashan DW Jr (2000) The relationship between serum IgE and surface levels of Fc epsilon R on human leukocytes in various diseases: correlation of expression with Fc epsilon RI on basophils but not on monocytes or eosinophils. J Allergy Clin Immunol 106(3):514–520

Sampson HA, Leung DY, Burks AW, Lack G, Bahna SL, Jones SM, Wong DA (2011) A phase II, randomized, doubleblind, parallelgroup, placebocontrolled oral food challenge trial of Xolair (omalizumab) in peanut allergy. J Allergy Clin Immunol 127(5):1309–1310. doi:10.1016/j.jaci. 2011.01.051 e1301

Schneider LC, Rachid R, Lebovidge J, Blood E, Mittal M, Umetsu DT (2013) A pilot study of omalizumab to facilitate rapid oral desensitization in high-risk peanut-allergic patients. J Allergy Clin Immunol 132(6):1368–1374. doi:10.1016/j.jaci.2013.09.046

Schroeder JT, Bieneman AP, Chichester KL, Hamilton RG, Xiao H, Saini SS, Liu MC (2010) Decreases in human dendritic cell-dependent T(H)2-like responses after acute in vivo IgE neutralization. J Allergy Clin Immunol 125(4):896–901. doi:10.1016/j.jaci.2009.10.021 e896

Sheinkopf LE, Rafi AW, Do LT, Katz RM, Klaustermeyer WB (2008) Efficacy of omalizumab in the treatment of atopic dermatitis: a pilot study. Allergy Asthma Proc 29(5):530–537. doi:10. 2500/aap.2008.29.3160

Soler M, Matz J, Townley R, Buhl R, O'Brien J, Fox H, Thirlwell J, Gupta N, Della Cioppa G (2001) The anti-IgE antibody omalizumab reduces exacerbations and steroid requirement in allergic asthmatics. Eur Respir J 18(2):254–261

Sutton BJ, Gould HJ (1993) The human IgE network. Nature 366(6454):421–428. doi:10.1038/ 366421a0

Sykes A, Edwards MR, Macintyre J, del Rosario A, Bakhsoliani E, Trujillo-Torralbo MB, Kon OM, Mallia P, McHale M, Johnston SL (2012) Rhinovirus 16-induced IFN-alpha and IFN-beta are deficient in bronchoalveolar lavage cells in asthmatic patients. J Allergy Clin Immunol 129 (6):1506–1514. doi:10.1016/j.jaci.2012.03.044 e1506

Xiong H, Curotto de Lafaille MA, Lafaille JJ (2012) What is unique about the IgE response? Adv Immunol 116:113–141. doi:10.1016/B978-0-12-394300-2.00004-1

Yamaguchi M, Lantz CS, Oettgen HC, Katona IM, Fleming T, Miyajima I, Kinet JP, Galli SJ (1997) IgE enhances mouse mast cell Fc(epsilon)RI expression in vitro and in vivo: evidence for a novel amplification mechanism in IgE-dependent reactions. J Exp Med 185(4):663–672

Yamamoto N, Takeshita K, Shichijo M, Kokubo T, Sato M, Nakashima K, Ishimori M, Nagai H, Li YF, Yura T, Bacon KB (2003) The orally available spleen tyrosine kinase inhibitor 2-[7-(3, 4-dimethoxyphenyl)-imidazo[1, 2-c]pyrimidin-5-ylamino]nicotinamide dihydrochloride (BAY 61-3606) blocks antigen-induced airway inflammation in rodents. J Pharmacol Exp Ther 306 (3):1174–1181. doi:10.1124/jpet.103.052316

New Insights on the Signaling and Function of the High-Affinity Receptor for IgE

Ryo Suzuki, Jörg Scheffel and Juan Rivera

Abstract Clustering of the high-affinity receptor for immunoglobulin E (FcεRI) through the interaction of receptor-bound immunoglobulin E (IgE) antibodies with their cognate antigen is required to couple IgE antibody production to cellular responses and physiological consequences. IgE-induced responses through FcεRI are well known to defend the host against certain infectious agents and to lead to unwanted allergic responses to normally innocuous substances. However, the cellular and/or physiological response of individuals that produce IgE antibodies may be markedly different and such antibodies (even to the same antigenic epitope) can differ in their antigen-binding affinity. How affinity variation in the interaction of FcεRI-bound IgE antibodies with antigen is interpreted into cellular responses and how the local environment may influence these responses is of interest. In this chapter, we focus on recent advances that begin to unravel how FcεRI distinguishes differences in the affinity of IgE–antigen interactions and how such discrimination along with surrounding environmental stimuli can shape the (patho) physiological response.

Abbreviations

BCR	B cell receptor
BTK	Bruton's tyrosine kinase
CBMC	Cord blood-derived mast cells
DAG	Diacylglycerol

The studies of the authors herein described were supported by the Intramural Research Program of the National Institute of Arthritis and Musculoskeletal and Skin Diseases of the National Institutes of Health.

R. Suzuki · J. Scheffel · J. Rivera
Molecular Immunology Section, Laboratory of Molecular Immunogenetics, National Institute of Arthritis and Musculoskeletal and Skin Diseases, National Institutes of Health, Bethesda, MD 20892, USA

J. Rivera (✉)
NIAMS-NIH, Building 10, Rm 13C103, Bethesda, MD 20892, USA
e-mail: juan_rivera@nih.gov

© Springer International Publishing Switzerland 2015
J.J. Lafaille and M.A. Curotto de Lafaille (eds.), *IgE Antibodies: Generation and Function*, Current Topics in Microbiology and Immunology 388,
DOI 10.1007/978-3-319-13725-4_4

DNP	Dinitrophenyl
ERK	Extracellular signal-regulated kinase
FcεRI	High-affinity receptor for immunoglobulin E
FRAP	Fluorescence recovery after photobleaching
FSMC	Fetal skin-derived mast cells
GAB2	GRB2-associated-binding protein 2
GFP	Green fluorescent protein
GIRKs	G-protein-coupled inwardly rectifying potassium channels
GM-CSF	Granulocyte macrophage colony-stimulating factor
HC	Highly cytokinergic
HSA	Human serum albumin
IgE	Immunoglobulin E
IP3	Inositol 1,4,5, triphosphate
ITAM	Immunoreceptor tyrosine-based activation motif
ITIM	Immunoreceptor tyrosine-based inhibitory motif
JNK	c-Jun N-terminal protein kinase
LAT1	Linker for activation of T cells
LAT2	Linker for activation of T cells family, member 2
LIF	Leukemia inhibitory factor
LTC4	Leukotriene C4
MAP Kinase	Mitogen-activated protein kinase
MCP-1	Monocyte chemoattractant protein-1
MIP-1	Macrophage inflammatory proteins-1
MITF	Microphthalmia transcription factor
NFAT	Nuclear factor of activated T cells
PC	Poorly cytokinergic
PCA	Passive cutaneous anaphylaxis
PI3-K	Phosphoinositide 3-kinase
PIP	Phosphatidylinositol 4,5 bisphosphate
PKC	Protein kinase C
PLC	Phospholipase C
PTK	Protein tyrosine kinase
QD	Quantum dot
RBL-2H3	Rat basophilic leukemia subclone-2H3
STAT	Signal transducer and activator of transcription
Syk	Spleen tyrosine kinase
TIRF	Total internal reflection fluorescence
TLR	Toll-like receptor
TNF	Tumor necrosis factor
TRB3	Tribbles homolog 3
VEGF	Vascular endothelial growth factor

Contents

1 Introduction 65
2 IgE Binding to FcεRI and Functional Consequences 67
   2.1 Does Monomeric IgE Binding to FcεRI Induce Mast Cell Responses? 67
   2.2 IgE-Antigen Engagement of FcεRI 69
3 Affinity of Receptor–Ligand Interactions and Functional Responses 72
   3.1 FcεRI and the Affinity of IgE/Antigen Interactions 73
   3.2 Deciphering the Affinity of IgE/Antigen Interactions by FcεRI 73
   3.3 In Vivo Consequences in Response to Differences in IgE/Antigen Affinities 75
4 Effect of the Microenvironment on FcεRI Responses 76
   4.1 Modulation of FcεRI Responses by G-Protein-Coupled Receptors 76
   4.2 Modulation of FcεRI Responses by Toll-like Receptors 81
5 Conclusion and Future Perspectives 82
References 83

1 Introduction

The binding of IgE antibodies to FcεRI occurs with high affinity and the half-life of receptor-bound IgE is measured in days (Isersky et al. 1979). When a multivalent antigen binds to the FcεRI-bound IgE antibodies, aggregation of individual receptors occurs (also known as receptor cross-linking); a key step in interpreting receptor engagement into downstream molecular signaling leading to the cells effector response (Metzger 1992). The molecular events that are initiated have been well reviewed (Gilfillan and Beaven 2011; Rivera et al. 2008). We will briefly summarize these events as background to the recent advances that will be detailed herein.

FcεRI is comprised of an IgE-binding α chain, a signal-amplifying β chain (Lin et al. 1996; On et al. 2004), and homodimeric γ chains (Gilfillan and Beaven 2011; Rivera et al. 2008). The β and γ chains encode immunoreceptor tyrosine-based activation motifs (ITAMs) (Pribluda et al. 1994), which are the targets for phosphorylation, a key step in receptor activation. Phosphorylation requires aggregation of FcεRI by IgE–antigen interactions, which allows proximity of receptors for transphosphorylation by the associated Src family protein tyrosine kinase (Src PTK), Lyn (Pribluda et al. 1994; Vonakis et al. 1997). Aggregation also segregates the clustered receptors in a cholesterol-enriched lipid domains in the plasma membrane, termed lipid rafts (Young et al. 2003), and protects the receptors from tyrosine phosphatases (Young et al. 2005), creating a phosphorylation-dominant environment that favors the phosphorylated form of FcεRI. The phosphorylated ITAM motifs act as docking modules that allow recruitment of SH2 domain-containing signaling molecules to FcεRI [such as Fyn and Syk kinase (Paolini et al. 1996; Parravicini et al. 2002; Yamashita et al. 2008)] which amplify the phosphorylation of this receptor and activate complementary signaling modules (pathways) that control the effector response of the cell (Nadler and Kinet 2002). Both protein and lipid molecules participate in this process (Olivera et al. 2007; Olivera and Rivera 2011; Oskeritzian et al. 2010), and their activity is mutually coordinated to initiate the appropriate cellular response (Fig. 1).

There are several known Src PTKs that participate in this process (Hong et al. 2007; Lee et al. 2011; Parravicini et al. 2002; Pribluda et al. 1994). In mast cells, it is known that Lyn is essential for phosphorylation of FcεRI (Pribluda et al. 1994), Fyn does not participate in FcεRI phosphorylation but controls a key lipid signaling pathway (the production of phosphatidyl-inositides) (Gomez et al. 2005a, b) that is essential in driving the effector responses of mast cells (Fig. 1). Other Src PTKs, such as Hck are known to negatively regulate that activity of Lyn kinase (Hong et al. 2007), whereas Fgr has been shown to be associated with FcεRI and presumably function as a positive regulator of mast cell effector responses (Lee et al. 2011), a topic that will be revisited below. Regardless, the implications are that FcεRI phosphorylation induces a complex series of events that are essential in the regulation of mast cell responses (Fig. 1). While for the most part, many of the signaling pathways and corresponding protein or lipid participants have been uncovered, it is not clear how these signaling molecules facilitate distinguishing a

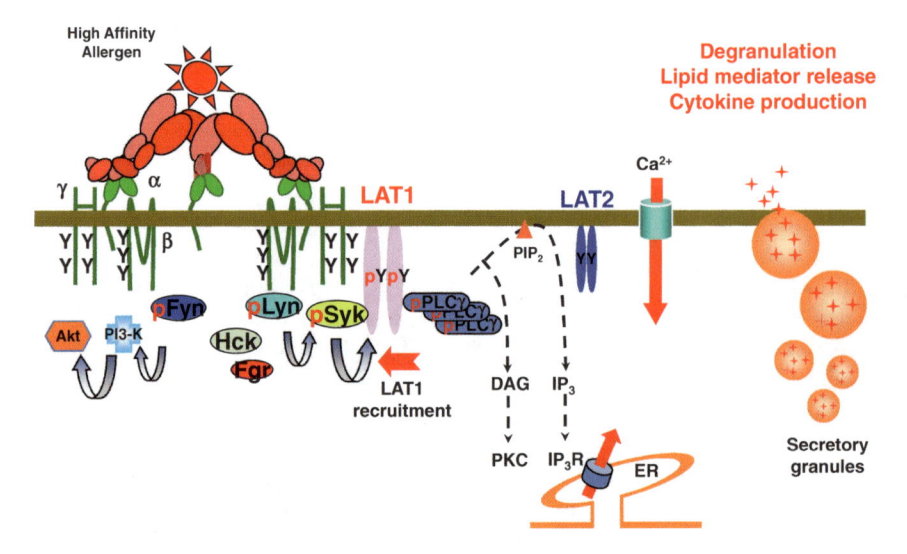

Fig. 1 Schematic model of FcεRI signaling by high-affinity IgE interactions with allergen. FcεRI is comprised of an IgE-binding α chain, a tretraspanning β chain, and two disulfide-linked γ chains. Aggregation of FcεRI with high-affinity allergen results in phosphorylation of the ITAMs by Lyn kinase and leads to activate Syk kinase through ITAM binding. Activation of Syk results in the phosphorylation of the adapter protein, linker of activation of T cells (LAT1), which is required for PLCγ phosphorylation, generation of IP$_3$, and normal calcium responses. The Lyn-Syk-LAT1 signals generated by high-affinity allergens cooperate with Fyn kinase signals that positively regulate PI3-K and Akt pathways. Together, these pathways are essential for the initiation of degranulation

strong from a weak stimulus. Moreover, these events occur in tissue-specific microenvironments where mast cells reside; thus, the influence of other factors that may be present or recruited to the tissues on FcεRI responses is an important consideration toward understanding the subtleties of how this receptors' engagement is ultimately interpreted into cellular responses. It is early in this endeavor; however, some recent advances are beginning to shed some light on such subtleties of IgE or IgE–antigen engagement of FcεRI.

2 IgE Binding to FcεRI and Functional Consequences

IgE binding to FcεRI occurs with very high affinity ($K_A \geq 10^{10} M^{-1}$) but it is reversible with a half-life of greater than six days (Furuichi et al. 1985). The high-affinity binding of IgE along with the apparent lack of immediate functional cellular consequences led to the theoretical and also practical view (particularly for experimental purposes) that the IgE-FcεRI complex comprises the "inert" or "resting" receptor (Metzger 1977, 1978). This was based on studies demonstrating that binding of IgE to cells in vitro failed to elicit the hallmark degranulation response of mast cells and only aggregation of the IgE with antigen could induce such response (Metzger 1983). Further support was provided by studies demonstrating a similar productive response (degranulation) of cells containing IgE and reacted with antibody to IgE and of cells that had not seen IgE but were stimulated with an aggregating antibody directed to FcεRI (Isersky et al. 1977, 1978). This suggested that the role of IgE was a passive one. Additional studies, using chemically oligomerized IgE, demonstrated that a response could be elicited by an IgE aggregate as small as a dimer, whereas monomers of IgE failed to provide a measurable response (Fewtrell and Metzger 1980). This suggested that dimerization of FcεRI is the minimal signal required for a measureable response.

2.1 Does Monomeric IgE Binding to FcεRI Induce Mast Cell Responses?

During the past decade, there has been considerable work demonstrating that monomeric IgE binding to FcεRI on mouse mast cells can induce various mast cells responses (Kalesnikoff et al. 2001; Yamaguchi et al. 1997) (Fig. 2). These include increased mast cell survival, increased cytokine production, and increased granule content and, in some cases, detectable molecular signals reviewed in (Kashiwakura et al. 2011; Kawakami and Galli 2002). One notable oddity, even in early studies, was that such responses seemed to be particular to the IgE clone used in the study. Later, it became evident that these clonal IgE molecules showed heterogeneity in their ability to induce responses. Highly cytokinergic (HC) IgEs (Kitaura et al. 2003, 2005)

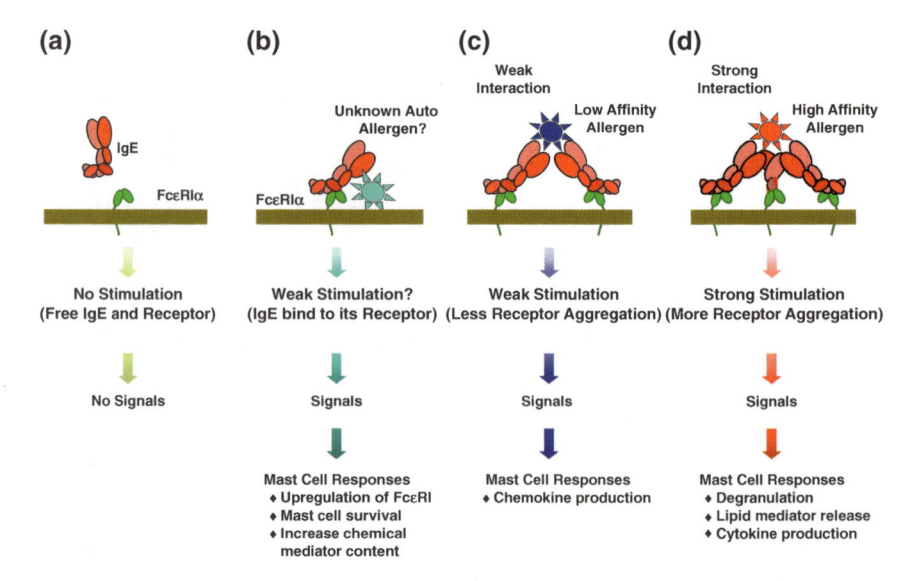

Fig. 2 Schematic summary of the varying stimulus strength and resulting effector response in mast cells. **a** A surface expressing high-affinity receptor for IgE (FcεRI) without IgE binding does not induce receptor activation. **b** Binding of monomeric IgE to FcεRI can cause cellular responses such as increasing cell surface expression of FcεRI and mast cell survival without effectively inducing degranulation and cytokine productions. This appears to occur by weak interactions of certain IgE's with unknown cell surface auto-antigens. **c** The weak interaction of a low-affinity allergen with cell surface-bound IgE antibody causes induction of cell signals that differ from those elicited by high-affinity antigens and results in induction of distinct responses (chemokine production, but not degranulation). **d** An aggregation of surface-bound IgE antibody with high-affinity allergen induces cellular signals that can effectively induce extensive mast cell effector responses (degranulation and cytokine productions)

promoted mast cell responses in the apparent absence of a multivalent antigen, and poorly cytokinergic (PC) IgEs did not show these responses. HC IgEs' also induced relatively strong signaling events with Lyn and Syk seemingly important to mast cell survival (Kitaura et al. 2005). PKCβII was implicated in the regulation of calcium responses to monomeric HC IgEs but less so in IgE-antigen-induced responses (Liu et al. 2005). Downstream molecules such as MAP kinases were also demonstrated to be active with HC IgEs but molecules such as Fyn, Gab2, PI3-K p85, or Akt were seemingly not required for monomeric IgE-mediated survival effects (Yamasaki et al. 2004). Nonetheless, other studies (Jimenez-Andrade et al. 2013) have found that monomeric IgE-induced VEGF secretion, through a Fyn kinase-dependent mechanism, modulates the activity of the translational regulator 4E-BP1 and that monomeric IgE increased melanoma tumor growth and blood vessel numbers in mouse. The induction of transcriptional activity by monomeric IgE has also been reported with activation of MITF, STAT3, and tribbles homolog 3 (TRB3) (Kuo et al. 2012; Sonnenblick et al. 2005).

It is still unclear as to whether monomeric HC IgE is relevant in humans. There is some evidence suggesting that human mast cells also respond to monomeric IgE. Gilchrest et al. (2003) reported that monomeric human IgE-induced chemokine I-309, GM-CSF, and MIP-1α from human cord blood-derived mast cells (CBMCs). Cruse et al. (2005) also demonstrated that monomeric IgE-induced Ca^{2+} influx, dose-dependent histamine release, LTC4 production, and IL-8 synthesis from human lung mast cells. Human CBMCs also produce IL-8 and MCP-1 in response to monomeric IgE (Matsuda et al. 2005). While these findings suggest that monomeric IgE may have previously unappreciated effects on cellular responses, there remains considerable ambiguity on whether these effects are indeed induced by a monomeric form of IgE or whether the IgEs inducing such effects have special properties that allow for FcεRI aggregation. Several pieces of evidence argue for this latter model: 1. Structural studies (James et al. 2003) have demonstrated that one type of mouse HC IgE (derived from the SPE-7 clone) changes its conformation upon binding to FcεRI inducing cross-reactivity to an undefined cell surface antigen (Fig. 2). 2. Most mouse HC IgEs have been shown to have polyreactivity to cellular autoantigens, and the functional effects of such IgEs can be disrupted by monovalent haptens (Kashiwakura et al. 2012). Based on such evidence, it appears that these HC IgEs are clustering (aggregating) FcεRI. Thus, in principle, the effects of "monomeric" IgE may differ from those of IgE-antigen clustering of the receptor by the extent of receptor clustering or perhaps the stability of such clusters. Regardless, these HC IgE's appear to maintain the requirement for FcεRI clustering to achieve cell activation and responses (Fig. 2). Whether some human IgE's share similar clustering characteristics remains unknown.

2.2 IgE-Antigen Engagement of FcεRI

2.2.1 Spatial Requirements for FcεRI Activation; Stability, and Mobility of This Receptor

Using trivalent DNP ligands with rigid DNA spacers of varying lengths (16, 26, 36, or 46 bases), Sil et al. (2007) demonstrated a relationship between the length of these ligands and the stimulation of FcεRI phosphorylation in RBL-2H3 cells. The findings showed that there is an inverse correlation with spacer length for FcεRI phosphorylation, with the shortest ligand being 5–6 fold stronger in stimulating this response than the longest ligand. In contrast, Ca^{2+} release from intracellular stores did not correlate with spacer length, suggesting that the spatial requirement was less stringent for this response and perhaps that the required signals were not spatially constrained or were separate from those required for FcεRI phosphorylation, as differences in PLCγ phosphorylation were not apparent. This suggests that the early events such as FcεRI phosphorylation are more spatially constrained than subsequent events. The use of a monovalent hapten (DNP-L-Lys), which diminishes

FcεRI aggregation by disrupting DNP-specific IgE interaction with a multivalent antigen (DNP$_{30-40}$-HSA), abrogates the phosphorylation of FcεRI and also causes inhibition of Ca^{2+} release from intracellular stores (Suzuki et al. 2013; Weetall et al. 1993). Thus, while Ca^{2+} release from intracellular stores is not spatially constrained, it appears to require that FcεRI remains aggregated in its appropriate spatial configuration for this response to occur. Simple dimerization alone does not appear to be sufficient, given that divalent ligands with similar DNA spacer lengths, as used for the trivalent ligands, failed to elicit responses and were effective blockers of multivalent antigen triggering of FcεRI (Paar et al. 2002).

Fluorescence recovery after photobleaching (FRAP) experiments from several different groups (Pyenta et al. 2003; Thomas et al. 1994) have shown that aggregation of IgE-FcεRI by multivalent antigen decreases the mobility of FcεRI clusters. The decrease in FcεRI cluster mobility appears to be associated with receptor activation. Carroll-Portillo et al. (2010) studied the engagement of DNP-specific IgE-occupied FcεRI with either DNP$_{24}$-BSA on lipid bilayers (mobile antigen) or DNP$_{24}$-BSA-conjugated coverslips (immobilized). Interaction of DNP-specific IgE-occupied FcεRI with immobilized DNP$_{24}$-BSA resulted in degranulation at the equivalent levels of mobile DNP$_{24}$-BSA. Although immobilized monovalent DNP-Lys failed to induce mast cell degranulation, unexpectedly, significant degranulation was observed under 10 or 25 mol % mobile monovalent DNP-embedded lipid bilayers; conditions that presumably do not cause receptor aggregation. The authors concluded that the FcεRI aggregation is not an obligatory step for mast cell signaling. However, studies in B cells have demonstrated that lipid bilayer presentation of monovalent antigens promotes changes in the B cell antigen receptor (BCR) that induces receptor aggregation (Tolar et al. 2009). Thus, whether presentation of monovalent antigen in a lipid bilayer to FcεRI promotes similar changes and receptor aggregation is currently not known.

The recent development of live-cell imaging using highly sensitive fluorescence microscopy and the improvement of fluorescent labeling techniques has revolutionized direct observation of the details of FcεRI cluster formation and mobility with high spatial and time resolution. Andrews et al. (2009) studied the relationship between antigen-induced changes in FcεRI mobility and signal initiation using total internal reflection fluorescence (TIRF) microscopy in combination with quantum dot (QD)-conjugated anti-DNP IgE (QD-IgE)-based single particle tracking. They found that the rate and extent of QD-IgE-FcεRI immobilization is dose-dependent in RBL-2H3 cells and in bone marrow-derived mast cells. Addition of low concentrations of a highly multivalent antigen (0.001–0.01 µg/ml DNP$_{25}$-BSA) caused little reduction in mobility of individual QD-IgE-FcεRI clusters and demonstrated they were relatively free to diffuse in the plasma membrane. However, high concentrations of antigen (1–10 µg/ml DNP$_{25}$-BSA) caused a marked reduction of movement of QD-IgE-FcεRI clusters. Antigen valencies were also found to affect FcεRI mobility. While the diffusion coefficient of FcεRI stimulated with DNP$_2$-BSA was indistinguishable from that of FcεRI in resting cells, a valency-associated reduction of FcεRI mobility was observed for DNP-HSA with valencies of 4, 12, and 25. Further, degranulation of mast cells was measured at the full range of doses and DNP-antigen valencies. Of note, degranulation occurred under conditions where receptors remained mobile. Immobile FcεRI appeared to be confined by actin-rich membrane-associated barriers, whereas mobile receptors (and diffusion) were largely limited to actin-poor regions of the membrane.

Thus, such studies indicate that a productive IgE/antigen engagement of FcεRI, leading to mast cell or basophil activation and effector function, is likely to require a spatially correct aggregation of FcεRI, which allows for effective transphosphorylation of this receptor (Sil et al. 2007). Downstream signaling is most productive when clusters of FcεRI are small and highly mobile (usually occurring at the periphery of the cell if antigen is immobilized in a lipid bilayer) and larger clusters of FcεRI are immobilized by actin-rich membrane barriers that appear to isolate these receptors from productive signaling modules or perhaps increase the dephosphorylation of FcεRI and thus limit signaling.

2.2.2 Antigen Concentration Effects

Varying the concentration of antigen to stimulate antigen-specific IgE antibodies bound to FcεRI has a marked effect on the cellular responses observed (degranulation, cytokine production, etc.). For example, suboptimal concentrations elicit very poor responses, which in some cases are not measurable. Supra-optimal concentrations also seem to dampen these responses. It has been suggested (Hlavacek et al. 2001) that the optimum concentration is one that sustains the presence of FcεRI in clusters for the time needed to enable the molecular events required for the mast cell effector responses. The kinetic proofreading model (McKeithan 1995), that is derived from studies on the T cell receptor (TCR), argues that receptors must remain clustered for a period of time sufficient to achieve the thresholds required for activation of downstream molecular signals. Experimental proof for such predictions can readily be obtained, but such studies are limited by selective focus on certain effector responses and do not necessarily encompass all cellular responses. Clues that warn against making the universal assumption that FcεRI must be clustered for prolonged periods to induce mast cell activation have begun to emerge.

Differences in the extent of FcεRI occupancy with antigen -specific IgE or in the concentration of antigen used to cluster FcεRI have demonstrated that changes in antigen concentration or in FcεRI occupancy with antigen-specific IgE can still induce responses in mast cells. Gonzalez-Espinosa et al. (2003) showed that varying the concentration of antigen elicited distinct profiles of cytokine/chemokine mRNA in murine bone marrow-derived mast cells. Quantitation of the net response to varying concentrations of antigen revealed that the mRNA accumulation of MIP-1α, MIP-1β, MCP-1, M-CSF, IL-2, and IL-4 reached a maximum at a significantly lower concentrations of antigen as compared to IL-3, IL-6, IL-10, LIF, MIP-2, TNF-α, and IFN-γ. Furthermore, differences in FcεRI occupancy with antigen-specific IgE (which may be similar to the state of an allergic individual whose mast cells may have varying amounts of allergen-specific IgE) showed that the degranulation response had linearity with increasing receptor occupancy. However, ~ 50 % of the maximal mRNA accumulation for MCP-1, M-CSF, and IL-4 was observed with ~ 10 % of receptor occupancy. Moreover, ~ 50 % of the maximal mRNA response for MIP-1α, MIP-1β, and MIP-2 required between 30 and 40 % IgE receptor occupancy. In contrast, IL-2, IL-6, IL-13, TNF-α, and LIF required greater than 40 % IgE receptor occupancy to achieve a 50 % response. Interestingly, the profile of cytokine/chemokine production seen at low receptor occupancy was accompanied by induction of p38 MAPK phosphorylation, whereas high receptor occupancy caused JNK1 and ERK2 activation. This demonstrated that sensitivity of signals differs and that selective induction of some molecular pathways may result in productive cellular responses since the secretion of some of these chemokines was detected (Gonzalez-Espinosa et al. 2003). Studies by Grodzki et al. (2009) also showed that antigen concentrations that are tenfold lower than those required to elicit FcεRI-dependent mast cell degranulation induced NFAT-mediated gene activation as measured by a GFP reporter assay. NFAT is essential for the generation of TNF-α and IL-13 in mast cells. NFAT activation was not blocked by a PI3-K inhibitor, whereas degranulation was inhibited by the treatment of the cells with this inhibitor. Thus, once again select and productive signals may be elicited by antigen concentrations that are likely to differ in their ability to sustain similar FcεRI clustering.

3 Affinity of Receptor–Ligand Interactions and Functional Responses

Studies in T cells assessing the effects of strong or weak ligands on TCR signaling demonstrated that weak stimuli generated unique signals/events when compared to strong ligands (Edwards and Evavold 2011; van der Merwe and Davis 2003). Surprisingly, the signaling behavior elicited by a weak stimulus was not mimicked by stimulation of the TCR with lower concentrations of strong ligands (Madrenas et al. 1995; Sloan-Lancaster et al. 1994). The fact that signals in response to weak stimulus are not identical to low concentrations of strong ligands suggested that a weak stimulus generates unique signals distinct from those that are evident upon the interaction of the TCR with a strong stimulus. The potential differences in molecular events elicited by a weak TCR stimulus are not well defined. Moreover, whether signaling elicited by a weak stimulus generates an effector response is unknown (Fig. 2).

3.1 FcεRI and the Affinity of IgE/Antigen Interactions

It is well known that distinct allergic individuals can produce IgE antibodies of differing affinities that recognize the same epitope of an allergy-producing antigen (termed allergen) (Jackola et al. 2002). How such differences in the affinity of IgE/antigen interactions are interpreted by FcεRI and what cellular outcomes may result has long remained an enigma (Fig. 2). Studies by Torigoe et al. (1998), began to explore this issue by investigating how antigens of differing affinities (approximately three orders of magnitude difference in affinity) affect mast cell signaling and degranulation. Engagement of DNP-specific IgE sensitized FcεRI on mast cells with low (2-nitrophenyl (2NP) caproate Fab; rapidly dissociating)- and high (dinitrophenyl (DNP) caproate Fab; slowly dissociating)-affinity antigens caused phosphorylation of this receptor with comparable amounts phosphorylated and with similar kinetics. High-affinity antigen was competent in transducing signals that resulted in mast cell degranulation (Suzuki et al. 2014). However, low-affinity antigen was ineffective in activating many of the same signaling molecules and failed to induce degranulation. Such findings are consistent with the aforementioned kinetic proofreading model (McKeithan 1995), arguing that the transduction of productive signals requires sustained clustering of FcεRI to produce the necessary levels of activated signaling molecules required for a productive response (Hlavacek et al. 2001). While such models may well explain how the threshold for producing a particular cellular response may be determined, they fail to consider that there are many cellular responses and some may not be limited by sustained FcεRI aggregation requirements. In addition, such models also fail to consider how the differences in the behavior of low-affinity versus high-affinity FcεRI clusters might be translated into subsequent molecular signals that may generate selective or unique responses as observed for the TCR (Edwards and Evavold 2011; van der Merwe and Davis 2003).

3.2 Deciphering the Affinity of IgE/Antigen Interactions by FcεRI

Using the same high- (DNP) and low- (2NP) affinity antigens used by Torigoe et al. (1998), we studied the effect of these antigens on the behavior of FcεRI clustering, how the differences in receptor clustering were interpreted into molecular signaling, and whether the generated molecular signals caused distinct cellular responses (Suzuki et al. 2014). To achieve parity in FcεRI phosphorylation a 100-fold greater concentration of low-affinity (2NP) was used relative to the high-affinity (DNP) antigen. Under these conditions, mast cell degranulation and cytokine production stimulated by the low-affinity antigen was poor relative to high-affinity antigen. Nonetheless, low-affinity antigen was a more potent stimulator of chemokine release compared to high-affinity antigen. To investigate the underlying mechanism(s) for this differential outcome under conditions where FcεRI was similarly phosphorylated, we studied the dynamics of the FcεRI clustering by TIRF microscopy and single particle tracking. Following high-affinity antigen (DNP) stimulation, small and highly mobile FcεRI clusters were observed whereas low-affinity (2NP) stimulation resulted in large and less mobile receptor clusters. Additionally, while high-affinity antigen induced a rapid mobilization of FcεRI clusters to the cell center forming a synapse-like structure, the low-affinity antigen showed a delayed movement of FcεRI clusters to the cell center, a loose formation of a synapse-like structure, and more of the FcεRI clusters remained at the cells periphery (Suzuki et al. 2014). Phosphotyrosine appeared to be more abundantly co-localized with the low-affinity FcεRI clusters although it was present in high-affinity FcεRI clusters as well. Perhaps, the most striking feature of low-affinity engagement observed by TIRF microscopy was the marked increase in association of FcεRI clusters with the Src PTK, Fgr. This kinase was implicated by others in FcεRI signaling and function upon high-affinity antigen engagement (Lee et al. 2011); however, we found only modest co-localization of Fgr with FcεRI under these conditions and in co-immunoprecipitation experiments endogenous Fgr was difficult to detect with FcεRI. In contrast, engagement of FcεRI with low-affinity antigen revealed detectable amounts of endogenous Fgr co-immunoprecipitating with FcεRI (Fig. 3). In addition, whereas high-affinity antigen failed to cause FcεRI phosphorylation in the absence of Lyn,

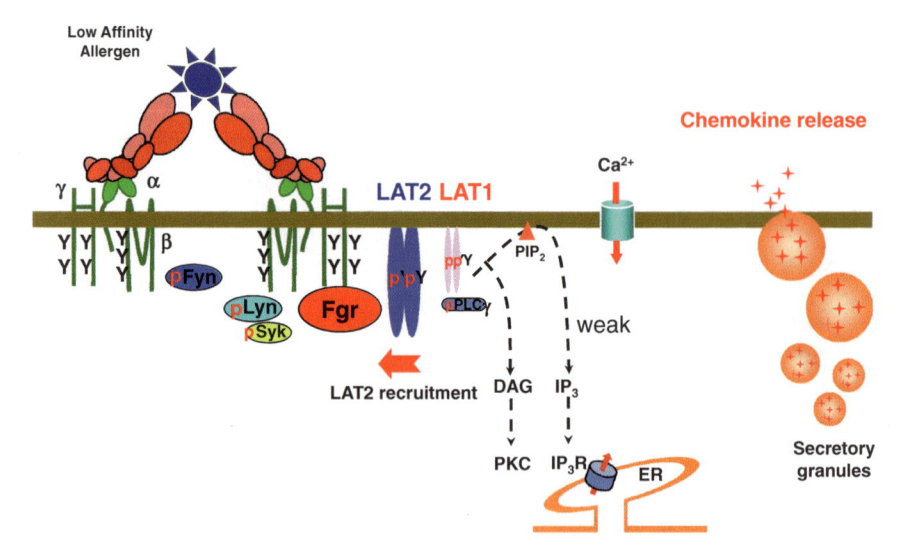

Fig. 3 Schematic model of FcεRI signals generated by low-affinity allergen. While the Lyn-Syk-LAT1 signals are dominant under high-affinity allergen engagement (see Fig. 1), low-affinity allergens are more effective in recruiting another Src kinase, Fgr, and cause enhanced colocalization of FcεRI clusters with the adapter LAT2, which is important for chemokine production. Thus, the lower allergen affinity causes changes in the molecular balance of both kinases (enhanced Fgr association) and adapter proteins (enhanced LAT2 co-localization) resulting in enhancement of chemokine release but dampened degranulation

low-affinity antigen caused a modest phosphorylation that was found to be Fgr-dependent (Suzuki et al. 2014). These findings demonstrate that FcεRI is able to distinguish low from high-affinity antigen engagement by shifting the balance of proximal Src PTKs used; a phenomena that seems to be related to the differences in FcεRI cluster size, mobility, and/or localization.

We further explored whether changes in the molecular balance of proximal signaling molecules might lead to differences in downstream signaling and alterations in cellular responses. Mast cell degranulation is tightly linked to the mobilization of calcium. Since degranulation was markedly reduced upon low-affinity stimulation of FcεRI, we investigated the calcium response to both low- and high-affinity antigens. The mobilization of calcium from intracellular stores as well as its influx from the extracellular medium was diminished when FcεRI was engaged by the low-affinity antigen as compared to high-affinity antigen. Exploration of the underlying mechanism revealed that high-affinity antigen engagement of FcεRI favored the phosphorylation of the adaptor LAT1 [an adaptor essential for calcium responses in mast cells (Saitoh et al. 2000)]. In contrast, low-affinity antigen engagement of FcεRI favored the phosphorylation of LAT2 (a LAT1-related adaptor), whose contribution to mast cell calcium responses and degranulation is modest, at best (Suzuki et al. 2014) (Fig. 3). These findings demonstrate that a distinct balance of signals results from low-affinity antigen engagement of FcεRI relative to high-affinity antigen engagement of this receptor. One concludes from such findings that the requirement for sustained FcεRI engagement (as proposed by the kinetic proofreading model) for a productive response is not absolute. Weak stimulation of FcεRI creates conditions that alter the molecular balance of signaling but such signals are still productive as they selectively enhance chemokine release (Figs. 2 and 3).

3.3 In Vivo Consequences in Response to Differences in IgE/Antigen Affinities

Of interest is, whether the changes in molecular signals and cellular responses caused by low-affinity antigen stimulation of FcεRI might result in differences in (patho) physiological responses. Using a model of passive cutaneous anaphylaxis (PCA), we tested this possibility by challenging the ears of mice sensitized with DNP-specific IgE with either high-affinity (DNP) or low-affinity (2NP) antigens. In both cases, an inflammatory response was induced although, remarkably, there was little mast cell degranulation in the ear tissue of mice challenged with low-affinity antigen (Suzuki et al. 2014). Nonetheless, much of the same symptomology was observed under either condition. However, analysis of the cell types in the ear tissue demonstrated a change in the profile of immune cell infiltrates; whereby high-affinity antigens induced a strong neutrophil response, low-affinity antigens induced increased numbers of monocyte/macrophages in ear tissues. This suggests the possibility that under low-affinity antigen conditions monocyte/macrophage recruitment may be required to induce a strong inflammatory response, given that these cells are more potent producers of cytokines/chemokines relative to neutrophils.

Collectively, the data discussed above suggests that differences in the strength of a stimulus can be deciphered by FcεRI. The clustering behavior, mobility, and localization of FcεRI appear to be involved in changes in the use of molecules that translate the affinity of the interaction into induction of selective signals and responses. This in turn has in vivo consequences with regard to the type of inflammatory response induced.

4 Effect of the Microenvironment on FcεRI Responses

Recently, there has been an increasing recognition of the effects of the microenvironment on FcεRI-induced cellular responses resulting from IgE/antigen interactions. The surrounding tissue microenvironment provides various soluble or membrane bound factors that can positively or negatively modulate IgE/antigen-activated mast cell responses. Multiple in vitro studies (as reviewed in Gilfillan and Beaven 2011) with mouse, rat or human mast cells or studies using in vivo models, support the concept that co-stimulation of other receptors together with FcεRI can significantly enhance, suppress, or modify mast cell responses induced through FcεRI. These receptors are mainly but not exclusively classified into five major categories (Table 1): (i) G-protein-coupled receptors (GPCRs), (ii) immunoreceptor tyrosine-based inhibitory motif (ITIM) (iii) activation motif (ITAM)-containing receptors, (iv) toll-like receptors (TLRs), and (v) cytokine receptors. Given the brevity of this chapter, we will discuss recent work focusing on GPCRs and TLRs (important contributors to innate immunity) as examples of how the tissue microenvironment may influence FcεRI signaling and function.

4.1 Modulation of FcεRI Responses by G-Protein-Coupled Receptors

Reports on modulation of IgE/antigen-dependent mast cell responses by GPCRs date back to 1978 when Marquardt and colleagues reported (Marquardt et al. 1978) an enhancement of mast cell responses to antigen challenge by adenosine—the endogenous ligand for P1 receptors. The P1 family consists of 4 members A1, A2A, A2B, and A3 which interacts with different heterotrimeric G-proteins, consisting of a G_α (with four different classes $G_{\alpha s}$, $G_{\alpha i/o}$, $G_{\alpha q/11}$, and $G_{\alpha 12/13}$) and $G_{\beta\gamma}$ subunits each of which are the product of multiple genes and splice variants. Like most GPCRs, the various P1 receptors interact with more than one G-protein, although they might have a preference for a particular one, but whose expression may vary among cells and with species. This promiscuity makes it difficult to

Table 1 Receptors and their respective ligands which modulate antigen-induced mast cell activation

Receptor family	Receptor on MCs	Ligand	Modulation through	Species tested	Effect on FcεRI-mediated activation	References
GPCR	S1P1	S1P	$G_{\alpha i/o}$	Mouse	Important for migration toward low antigen	Jolly et al. (2004), Olivera (2008)
	S1P2		$G_{\alpha i/o}, G_{\alpha q/11}, G_{\alpha 12/13}$	Mouse	Enhanced degranulation	Jolly et al. (2004), Olivera (2008)
	CCR1	CCL3 (MIP-1α), CCL5 (RANTES)	$G_{\alpha i/o}, G_{\alpha q/11}, G_{\alpha 12/13}$	Rat/mouse	Enhanced degranulation, cytokine release	Laffargue et al. (2002), Miyazaki et al. (2005), Toda et al. (2004)
	CCR3	CCL11	$G_{\alpha i/o}$	Human	Enhanced IL-13 release	Price et al. (2002)
	A2A	Adenosine	$G_{\alpha s}$	Human	Time-dependent inhibited or enhanced degranulation (partially through uptake via ENT1)	Hughes et al. (1984), Gomez et al. (2013)
	A2B		$G_{\alpha s}, G_{\alpha q/11}$	Human	Enhanced cytokine release	Feoktistov and Biaggioni (1995), Ryzhov et al. (2006)
	A3		$G_{\alpha i/o}, G_{\alpha q/11}, G_{\alpha 12/13}$	Rat/mouse	Enhanced degranulation	Laffargue et al. (2002), Ramkumar et al. (1993)
	EP1	PGE2	$G_{\alpha q/11}$	Human	Enhanced degranulation	Feng et al. (2006), Wang and Lau (2006)
	EP2		$G_{\alpha s}$	Human	Inhibited degranulation and cytokine release	Feng et al. (2006), Serra-Pages et al. (2012)
	EP3		$G_{\alpha i/o}, G_{\alpha q/11}, G_{\alpha s}$	Human/mouse	Enhanced degranulation	Nguyen et al. (2002), Serra-Pages et al. (2012)
	EP4		$G_{\alpha i/o}, G_{\alpha q/11}, G_{\alpha s}$	Human	Enhanced cytokine release, probably inhibitory in human	Feng et al. (2006)
	Melanocortin-1 (MC-1)	α-MSH	$G_{\alpha s}$	Mouse	Inhibited degranulation and selected cytokines	Adachi et al. (1999)
	Neurokinin-1	Substance P	$G_{\alpha i/o}, G_{\alpha q/11}$ /decreased FcεRI expression	Rat/human	Enhanced degranulation /inhibited leukotriene production (24 h pre-treatment)	Liao et al. (2006), McCary et al. (2010)
	P2Y1 (potentially other P2Y and P2X)	ATP, ADP, UTP	$G_{\alpha q/11}, G_{\alpha 12/13}$	Human	Enhanced degranulation	Schulman et al. (1999), Gao et al. (2012)
	C3aR	C3a	$G_{\alpha i/o}$	Human (not expressed on RBL or mouse MC)	Enhanced FcγR-induced degranulation (potentially FcεRI as well)	Woolhiser et al. (2004)
	Independent of C3aR acting on FcεRI itself	C3a	Interference with β-chain assembly	Rat/mouse	Inhibited degranulation	Erdei et al. (1995, 1999)
	β2 adrenergic receptor	β2 agonists	$G_{\alpha s}$	Human	Inhibited degranulation and cytokine production	Bissonnette and Befus (1997)

(continued)

Table 1 (continued)

Receptor family	Receptor on MCs	Ligand	Modulation through	Species tested	Effect on FcεRI-mediated activation	References
ITIM-bearing receptors	FcγRIIB	IgG-complexes	SHIP-1	Rat (transfected)	Inhibited degranulation and cytokine release	Daeron et al. (1995)
	Allergin-1	n.d.	SHP-1, SHP-2	Mouse	Inhibited degranulation	Hitomi et al. (2010)
	gp49B1	Integrin αvβ3	SHIP-1	Mouse	Inhibited degranulation and leukotriene production	Castells et al. (2001), Katz et al. (1996)
	SIRP-α (CD172a)	CD47/IAP	SHP-1, SHP-2	Rat (human)	Inhibited serotonin and TNFα release, decreased Ca^{2+} responses	Lienard et al. (1999)
	CD300a (MAIR-1, Irp60, LMIR1, CMRF-35)	Phosphatidylethanolamine, Phosphatidylserine	SHP-1, SHP-2	Mouse /human	Inhibited degranulation and cytokine release	Bachelet et al. (2005), Yotsumoto et al. (2003)
	CD300f (LMIR3)	Ceramide	SHP-1, SHP-2	Mouse	Inhibited cytokine release	Izawa et al. (2009), Izawa et al. (2012)
	CD305 (LAIR-1)	Collagens	SHP-1, SHP-2	Rat	Inhibited degranulation	Meyaard (2008)
	Siglec 8	Sialylated structures	n.d.	Human	Inhibited degranulation and leukotriene production	Yokoi et al. (2008)
	MAFA	n.d.	SHP-2	Rat (not expressed on human or mouse MC)	Inhibited degranulation and cytokine release	Guthmann et al. (1995), Ortega Soto and Pecht (1988)
	CD31 (PECAM-1)	Fibronectin, PECAM-1, CD38, integrin αvβ3	SHP-2	Mouse	Deficiency causes enhanced serotonin release in vitro and enhanced anaphylaxis in vivo	Wong et al. (2002)
	PIR-B (LIR-3, LILRB-3)	MHC-1	Independent of SHP-1, SHP-2 and SHIP-1	Rat	Inhibited mast cell degranulation	Uehara et al. (2001)
	TLT-1	n.d.	SHP-2	Rat (transfected)	Enhanced Ca^{2+} signaling	Barrow et al. (2004)
ITAM-bearing receptors	FcαRI (CD89)	Monomeric IgA	ITAM	Rat	Inhibited degranulation	Pasquier et al. (2005)
	FcαRI	IgA-complexes	ITAM	Rat	Enhanced mast cell activation	Pasquier et al. (2005)
	FcγRI	IgG-complexes	ITAM	Human	IFNγ-sensitized cells show enhanced degranulation	Woolhiser et al. (2001)

(continued)

Table 1 (continued)

Receptor family	Receptor on MCs	Ligand	Modulation through	Species tested	Effect on FcεRI-mediated activation	References
Toll-like receptors	TLR1/TLR2	PAM₃CSK₄	Various potential sites	Human/mouse	Inhibited degranulation, enhanced cytokine release	Fehrenbach et al. (2007), Kasakura et al. (2009)
	TLR6/TLR2	MALP-2	Various potential sites	Mouse	Enhanced cytokine production	Fehrenbach et al. (2007), Qiao et al. (2006)
	TLR4	LPS, ES-62 (other ligands not tested)	Various potential sites	Mouse	Enhanced degranulation and cytokine release	Melendez et al. (2007), Qiao et al. (2006)
Cytokine receptors	IL-3R	IL-3	JAK-STAT	Human	Enhanced histamine and leukotriene release	Gebhardt et al. (2002)
	IL-4R	IL-4	JAK-STAT /decreased FcεRI expression	Human	Enhanced cytokine and leukotriene release/inhibited cytokines	Hsieh et al. (2001), Toru et al. (1998)
	IL-5R	IL-5	JAK-STAT	Human	Enhanced cytokine release	Ochi et al. (2000)
	IL-10R	IL-10	JAK-STAT	Human/mouse	Inhibited cytokine release and FcεRIβ expression	Arock et al. (1996), Kennedy Norton et al. (2008)
	CD117 (c-Kit)	SCF	Receptor tyrosine kinase	Mouse/human	Enhanced degranulation and cytokine release	Ishizuka et al. (1998), Tkaczyk et al. (2004)
	TGFBR1, -R2, -R3	TGF-β	Receptor serin/threonin kinase, decreased FcεRI expression	Human/mouse/rat	Inhibited degranulation and cytokine release	Bissonnette et al. (1997), Gomez et al. (2005b)
Others	4-1BB (TNFRSF9)	4-1BB ligand	Potentially via Lyn	Mouse	4-1BB deficient MC have decreased degranulation, and cytokine release	Nishimoto et al. (2005)
	CD28	B7	n.d.	Mouse	Enhanced TNFα release	Tashiro et al. (1997)
	CD200R	CD200 (OX2)	NPXY motif, Dok1-SHIP, Dok2-RasGAP	Human/mouse	Inhibited degranulation and cytokine release	Cherwinski et al. (2005), Zhang et al. (2004)
	CD226 (DNAM-1, PTA1)	CD112 (Nectin-2), CD155	Fyn	Human	Enhanced degranulation and prostaglandin release	Bachelet et al. (2006)
	CD252 (OX40L, TNFSF4)	CD134 (OX40, TNFRSF4)	Increased cAMP	Mouse	Inhibited degranulation	Gri et al. (2008), Sibilano et al. (2012)
	TIM-1, TIM-3	Phosphatidylserine	Tyrosine phosphorylation motive	Mouse	Enhanced cytokines production	Nakae et al. (2007)
	RAGE	S100I2A	n.d.	Human/mouse	Enhanced Histamin and leukotrien release	Yang et al. (2007)

n.d. not determined, *α-MSH* alpha melanocyte-stimulating hormone, *CCL/R* chemokine C–C motive ligand/receptor, *Dok1* docking protein, *GPCR* G-protein-coupled receptor, *LPS* lipopolysaccharide, *MAFA* mast cell function-associated antigen, *MHC* major histocompatibility complex, *PECAM-1* platelet endothelial cell adhesion molecule, *PGE2* prostaglandin E2, *PIR-B* paired immunoglobulin-like receptor B, *RAGE* receptor for advanced glycation end products, *S1P1/2* sphingosine receptor ½, *SHIP SH2* domain-containing inositol 5′-phosphatase, *SHP SH2* domain-containing phosphatase, Siglec sialic acid-binding lectin, Ig-like lectin, *SIRP-α* signal-regulatory protein alpha, *TGF-β* transforming growth factor-β, *TIM-1/2* T-cell immunoglobulin and mucin domain, *TLR* toll-like receptor, *TLT-1* TREM-like transcript-1, and *TNFRSF* tumor necrosis factor receptor superfamily predict the effect of a particular GPCR on FcεRI signaling and responses in mast cells. However, in general, $G_{\alpha s}$ has been found to stimulate adenyl cyclases which catalyze the conversion of ATP to 3′,5′-cyclic AMP (cAMP) and pyrophosphate while $G_{\alpha i}$ mainly inhibits them. Although it's not entirely understood how cAMP acts in mast cells, it seems that elevated cAMP levels dampens mast cell calcium responses and degranulation (reviewed in Alm and Bloom 1982). Consistent with this view A2A and A2B receptors, which strongly interact with $G_{\alpha s}$, suppress FcεRI-induced mast cell degranulation, while A3 receptors which associate with $G_{\alpha i}$ potentiate it (Table 1). Such examples extend beyond the adenosine receptor family, as recent studies by Serra-Pages et al. (2012) showed that prostaglandin E_2 (PGE$_2$) can enhance or suppress FcεRI-dependent mast cell degranulation depending on the ratio of EP$_2$ to EP$_3$ receptor expression on these cells, which, respectively, associate with $G_{\alpha s}$ or $G_{\alpha i}$. Interestingly, negative regulation of mast cell degranulation through cAMP is not restricted to GPCRs. In a study focusing on the modulating capacity of the mast cell-expressed OX40L, Gri et al. (2008) demonstrated that Foxp3$^+$T regulatory cells were able to dampen the release of granule-stored mediators from mast cells through OX40L-OX40 interaction which led to increased intracellular cAMP in BMMC. The GPCRs that enhance FcεRI mast cell responses are likely to do so via multiple mechanisms. $G_{\alpha q/11}$-associating receptors activate PLC to catalyze the cleavage of phosphatidylinositol 4,5-biphosphate (PIP$_2$) into inositol (1,4,5) triphosphate (IP$_3$) and diacylglycerol (DAG) which activates intracellular Ca^{2+} stores via IP$_3$ receptors or the PKC pathway, respectively, providing stronger calcium responses linked to enhanced mast cell degranulation. Furthermore, they activate brutons tyrosine kinase (BTK) which is an important signaling molecule in FcεRI receptor signaling and helps to regulate calcium (reviewed in Kawakami et al. 1999). $G_{\alpha 12/13}$-coupled GPCRs stimulate the small GTPase Rho, Na$^+$/H$^+$ exchangers, and JNK. The GTPase Rho is involved in the regulation of mast cell degranulation (Price et al. 1995; Sibilano et al. 2012). Accordingly, GPCRs that associate with $G_{\alpha q11}$ and/or $G_{\alpha 12/13}$, such as A2B and A3, the chemokine receptor CCR1 or the receptors for PGE2; EP1, EP3, and EP4, usually enhance FcεRI signaling and potentiate degranulation and/or cytokine release (Table 1). Although probably most of the observed effects of GPCRs on mast cell responses can be explained by the activity of the G_α subunit, it should be taken into account that $G_{\beta\gamma}$ by itself regulates several signaling processes such as PI3-K, adenylylcyclases, PLCβ, BTK, or G-protein-coupled inwardly rectifying potassium channels (GIRKs) and thereby can modulate G_α signals (reviewed in Birnbaumer 2007). Therefore, the association of a particular G_α protein with a given receptor should not be viewed as entirely predictive of whether the modulatory role might be enhancing or suppressing.

4.2 Modulation of FcεRI Responses by Toll-like Receptors

Present knowledge suggests that rodent and human mast cells or basophils can express a variety of TLRs including 1, 2, 3, 4, 5, 6, 7, and 9, which can activate these cells independently but also modulate FcεRI signaling and responses (Komiya et al. 2006; Novak et al. 2010). However, expression of TLRs appears to vary. On mast cells, the difference in expression is related to the species or tissue source from which they were isolated or, for in vitro cultured cells, the cytokine environment in which they are cultured. For instance, TLR3, 7, and 9 ligands failed to stimulate mouse BMMC due to the low expression of these receptors, whereas murine fetal skin-derived-cultured mast cells (FSMC) responded by expressing higher levels of mRNA for cyto- and chemokines (Matsushima et al. 2004). Some cultured human mast cells apparently lack TLR4 expression but it can be upregulated and functionality restored by exposure to IL-4 or IFN-γ (Okumura et al. 2003; Varadaradjalou et al. 2003). Moreover, rat peritoneal mast cells can upregulate TLR2 and TLR4 under the influence of CCL5 and/or IL-6 (Pietrzak et al. 2011). Thus, such findings suggest that the cytokine milieu influences which microbial or endogenous pathogen-associated molecular patterns can be recognized and directly stimulate mast cells or alter FcεRI-dependent activation. Interestingly, although TLR stimulation by itself activates mast cells, co-stimulation of TLRs and FcεRI is not per se synergistic. Instead, the outcome may be determined by the ligand activating a given TLR or combination of receptors. Fehrenbach et al. (2007) showed that the TLR1/2 ligand Pam$_3$CSK$_4$ attenuated IgE-antigen-induced degranulation and Ca^{2+} mobilization of mouse BMMCs in a TLR2-dependent manner, while the TLR6/2 ligand MLAP-2 did not. However, both ligands synergized with FcεRI signals to induce IL-6. The inhibitory effect of Pam$_3$CSK$_4$ was also confirmed in vivo (Kasakura et al. 2009), while the enhanced IL-6 release through TLR2-FcεRI co-stimulation could be induced with extracts from *M. sympodialis*, a skin commensal yeast (Selander et al. 2009). In addition, Yoshioka et al. (2007) reported that lipoteichoic acid (LTA) and peptidoglycan—both TLR2 agonists—dampened IgE-induced degranulation of LAD2 and human pulmonary mast cells. This inhibitory effect appeared to be achieved through downregulation of FcεRI surface expression. TLR2 is not the only TLR found to negatively impact mast cell responses to FcεRI stimulation as ES-62—a secreted product of filarial nematodes that binds to TLR4—was found to inhibit antigen-induced responses in human mast cells by interfering with FcεRI signaling events through presumed degradation of PKC-α (Melendez et al. 2007). In contrast, stimulation of TLR4 through LPS generally augments FcεRI-dependent mast cell responses (Table 1). How these differences in modulation of FcεRI responses by a given TLR arise (engaged by distinct ligands) is not understood. Nonetheless, there are many potential intersection points in TLR and FcεRI signaling cascades (Novak et al. 2010), and there is evidence that TLRs may utilize some of the same key signaling molecules used by FcεRI, such as Syk, Lyn, or Btk (Avila et al. 2012; Chaudhary et al. 2007; Horwood et al. 2006). However, some mast cell-specific peculiarities may also influence the role of TLRs in modulating FcεRI-dependent responses. For instance, mast cells do not express CD14, an important co-receptor for the recognition of LPS and other TLR4 ligands (Bischoff 2007). Additionally, the use or expression of certain signaling molecules may differ in mast cells as compared to other cell types. For example, Zorn et al. (2009) demonstrated that Btk is dispensable for TLR4- or TLR1/2-dependent activation of mouse BMMC with LPS or Pam$_3$CSK$_4$, respectively, but yet it played a significant role in TLR signaling in other cell types (Horwood et al. 2006; Jefferies et al. 2003).

Because TLRs belong to the family of pattern recognition receptors, it should be recognized that each TLR recognizes several molecules that share conserved molecular patterns. The evidence to date suggests that distinct TLR ligands can generate distinct responses through the same TLR. Thus, to uncover the molecular mechanism by which TLRs modulate FcεRI signaling and responses, one must decipher the molecular signals of each TLR and the respective ligand regarding its influence on FcεRI signaling. Uncovering of these complexities is key toward understanding how the microenvironment exerts its influence on FcεRI-dependent mast cell responses.

5 Conclusion and Future Perspectives

Current knowledge on how FcεRI interprets the interaction of receptor-bound IgE antibodies with cognate antigen is still evolving. The recent advances outlined above shed new light on the molecular machinery utilized by FcεRI and how it can be influenced by the surrounding environmental milieu. This new information reveals important subtleties in how IgE antibodies are able to elicit discrete signals through FcεRI that promote distinct effector responses with physiological consequences. Noteworthy is the recognition that differences in the affinity of IgE antibody interactions with antigen can be interpreted by FcεRI, resulting in changes in the immune response. Of similar note is the increasing evidence that IgE-dependent responses through FcεRI can also be modulated by the microenvironment in which mast cells or basophils may be found, through the plethora of receptors expressed on these cells. Some of these co-modulators can dampen some effector responses (degranulation) while enhancing others (cytokine production). This raises the possibility that IgE-dependent responses might be modulated for therapeutic purposes through identification of appropriate molecular targets. Future efforts might be focused on deciphering the intersecting molecular signals engendered by these various modulators and whether there is efficacy in specifically tailoring FcεRI responses in vivo. Regardless, the recent work clearly demonstrates that the biology and function of IgE-dependent FcεRI responses is highly complex and much remains to be revealed.

References

Adachi S, Nakano T, Vliagoftis H, Metcalfe DD (1999) Receptor-mediated modulation of murine mast cell function by alpha-melanocyte stimulating hormone. J Immunol 163(6):3363–3368

Alm E, Bloom GD (1982) Cyclic nucleotide involvement in histamine release from mast cells–a reevaluation. Life Sci 30(3):213–218

Andrews NL, Pfeiffer JR, Martinez AM, Haaland DM, Davis RW, Kawakami T, Oliver JM, Wilson BS, Lidke DS (2009) Small, mobile FcεRI receptor aggregates are signaling competent. Immunity 31(3):469–479

Arock M, Zuany-Amorim C, Singer M, Benhamou M, Pretolani M (1996) Interleukin-10 inhibits cytokine generation from mast cells. Eur J Immunol 26(1):166–170

Avila M, Martinez-Juarez A, Ibarra-Sanchez A, Gonzalez-Espinosa C (2012) Lyn kinase controls TLR4-dependent IKK and MAPK activation modulating the activity of TRAF-6/TAK-1 protein complex in mast cells. Innate Immun 18(4):648–660

Bachelet I, Munitz A, Mankutad D, Levi-Schaffer F (2006) Mast cell costimulation by CD226/CD112 (DNAM-1/Nectin-2): a novel interface in the allergic process. J Biol Chem 281 (37):27190–27196

Bachelet I, Munitz A, Moretta A, Moretta L, Levi-Schaffer F (2005) The inhibitory receptor IRp60 (CD300a) is expressed and functional on human mast cells. J Immunol 175(12):7989–7995

Barrow AD, Astoul E, Floto A, Brooke G, Relou IA, Jennings NS, Smith KG, Ouwehand W, Farndale RW, Alexander DR, Trowsdale J (2004) Cutting edge: TREM-like transcript-1, a platelet immunoreceptor tyrosine-based inhibition motif encoding costimulatory immunoreceptor that enhances, rather than inhibits, calcium signaling via SHP-2. J Immunol 172 (10):5838–5842

Birnbaumer L (2007) Expansion of signal transduction by G proteins. The second 15 years or so: from 3 to 16 alpha subunits plus betagamma dimers. Biochim Biophys Acta 1768(4):772–793

Bischoff SC (2007) Role of mast cells in allergic and non-allergic immune responses: comparison of human and murine data. Nat Rev Immunol 7(2):93–104

Bissonnette EY, Befus AD (1997) Anti-inflammatory effect of β 2-agonists: inhibition of TNF-α release from human mast cells. J Allergy Clin Immunol 100(6 Pt 1):825–831

Bissonnette EY, Enciso JA, Befus AD (1997) TGF-β1 inhibits the release of histamine and tumor necrosis factor-α from mast cells through an autocrine pathway. Am J Respir Cell Mol Biol 16 (3):275–282

Carroll-Portillo A, Spendier K, Pfeiffer J, Griffiths G, Li H, Lidke KA, Oliver JM, Lidke DS, Thomas JL, Wilson BS, Timlin JA (2010) Formation of a mast cell synapse: FcεRI membrane dynamics upon binding mobile or immobilized ligands on surfaces. J Immunol 184(3):1328–1338

Castells MC, Klickstein LB, Hassani K, Cumplido JA, Lacouture ME, Austen KF, Katz HR (2001) gp49B1-α(v)β3 interaction inhibits antigen-induced mast cell activation. Nat Immunol 2 (5):436–442

Chaudhary A, Fresquez TM, Naranjo MJ (2007) Tyrosine kinase Syk associates with toll-like receptor 4 and regulates signaling in human monocytic cells. Immunol Cell Biol 85(3):249–256

Cherwinski HM, Murphy CA, Joyce BL, Bigler ME, Song YS, Zurawski SM, Moshrefi MM, Gorman DM, Miller KL, Zhang S, Sedgwick JD, Phillips JH (2005) The CD200 receptor is a novel and potent regulator of murine and human mast cell function. J Immunol 174(3):1348–1356

Cruse G, Kaur D, Yang W, Duffy SM, Brightling CE, Bradding P (2005) Activation of human lung mast cells by monomeric immunoglobulin E. Eur Respir J 25(5):858–863

Daeron M, Malbec O, Latour S, Arock M, Fridman WH (1995) Regulation of high-affinity IgE receptor-mediated mast cell activation by murine low-affinity IgG receptors. J Clin Invest 95 (2):577–585

Edwards LJ, Evavold BD (2011) T cell recognition of weak ligands: roles of signaling, receptor number, and affinity. Immunol Res 50(1):39–48

Erdei A, Andreev S, Pecht I (1995) Complement peptide C3a inhibits IgE-mediated triggering of rat mucosal mast cells. Int Immunol 7(9):1433–1439

Erdei A, Toth GK, Andrasfalvy M, Matko J, Bene L, Bajtay Z, Ischenko A, Rong X, Pecht I (1999) Inhibition of IgE-mediated triggering of mast cells by complement-derived peptides interacting with the Fc ε RI. Immunol Lett 68(1):79–82

Fehrenbach K, Port F, Grochowy G, Kalis C, Bessler W, Galanos C, Krystal G, Freudenberg M, Huber M (2007) Stimulation of mast cells via FcvarepsilonR1 and TLR2: the type of ligand determines the outcome. Mol Immunol 44(8):2087–2094

Feng C, Beller EM, Bagga S, Boyce JA (2006) Human mast cells express multiple EP receptors for prostaglandin E2 that differentially modulate activation responses. Blood 107(8):3243–3250

Feoktistov I, Biaggioni I (1995) Adenosine A2b receptors evoke interleukin-8 secretion in human mast cells. An enprofylline-sensitive mechanism with implications for asthma. J Clin Invest 96(4):1979–1986

Fewtrell C, Metzger H (1980) Larger oligomers of IgE are more effective than dimers in stimulating rat basophilic leukemia cells. J Immunol 125(2):701–710

Furuichi K, Rivera J, Isersky C (1985) The receptor for immunoglobulin E on rat basophilic leukemia cells: effect of ligand binding on receptor expression. Proc Natl Acad Sci USA 82(5):1522–1525

Gao ZG, Wei Q, Jayasekara MP, Jacobson KA (2012) The role of P2Y(14) and other P2Y receptors in degranulation of human LAD2 mast cells. Purinergic Signal 9(1):31–40

Gebhardt T, Sellge G, Lorentz A, Raab R, Manns MP, Bischoff SC (2002) Cultured human intestinal mast cells express functional IL-3 receptors and respond to IL-3 by enhancing growth and IgE receptor-dependent mediator release. Eur J Immunol 32(8):2308–2316

Gilchrest H, Cheewatrakoolpong B, Billah M, Egan RW, Anthes JC, Greenfeder S (2003) Human cord blood-derived mast cells synthesize and release I-309 in response to IgE. Life Sci 73(20):2571–2581

Gilfillan AM, Beaven MA (2011) Regulation of mast cell responses in health and disease. Crit Rev Immunol 31(6):475–529

Gomez G, Gonzalez-Espinosa C, Odom S, Baez G, Cid ME, Ryan JJ, Rivera J (2005a) Impaired FcεRI-dependent gene expression and defective eicosanoid and cytokine production as a consequence of Fyn deficiency in mast cells. J Immunol 175(11):7602–7610

Gomez G, Ramirez CD, Rivera J, Patel M, Norozian F, Wright HV, Kashyap MV, Barnstein BO, Fischer-Stenger K, Schwartz LB, Kepley CL, Ryan JJ (2005b) TGF-β 1 inhibits mast cell Fc ε RI expression. J Immunol 174(10):5987–5993

Gomez G, Nardone V, Lotfi-Emran S, Zhao W, Schwartz LB (2013) Intracellular adenosine inhibits IgE-dependent degranulation of human skin mast cells. J Clin Immunol 33(8):1349–1359

Gonzalez-Espinosa C, Odom S, Olivera A, Hobson JP, Martinez ME, Oliveira-Dos-Santos A, Barra L, Spiegel S, Penninger JM, Rivera J (2003) Preferential signaling and induction of allergy-promoting lymphokines upon weak stimulation of the high affinity IgE receptor on mast cells. J Exp Med 197(11):1453–1465

Gri G, Piconese S, Frossi B, Manfroi V, Merluzzi S, Tripodo C, Viola A, Odom S, Rivera J, Colombo MP, Pucillo CE (2008) CD4 + CD25 + regulatory T cells suppress mast cell degranulation and allergic responses through OX40-OX40L interaction. Immunity 29(5):771–781

Grodzki AC, Moon KD, Berenstein EH, Siraganian RP (2009) FcεRI-induced activation by low antigen concentrations results in nuclear signals in the absence of degranulation. Mol Immunol 46(13):2539–2547

Guthmann MD, Tal M, Pecht I (1995) A new member of the C-type lectin family is a modulator of the mast cell secretory response. Int Arch Allergy Immunol 107(1–3):82–86

Hitomi K, Tahara-Hanaoka S, Someya S, Fujiki A, Tada H, Sugiyama T, Shibayama S, Shibuya K, Shibuya A (2010) An immunoglobulin-like receptor, Allergin-1, inhibits immunoglobulin E-mediated immediate hypersensitivity reactions. Nat Immunol 11(7):601–607

Hlavacek WS, Redondo A, Metzger H, Wofsy C, Goldstein B (2001) Kinetic proofreading models for cell signaling predict ways to escape kinetic proofreading. Proc Natl Acad Sci USA 98 (13):7295–7300

Hong H, Kitaura J, Xiao W, Horejsi V, Ra C, Lowell CA, Kawakami Y, Kawakami T (2007) The Src family kinase Hck regulates mast cell activation by suppressing an inhibitory Src family kinase Lyn. Blood 110(7):2511–2519

Horwood NJ, Page TH, McDaid JP, Palmer CD, Campbell J, Mahon T, Brennan FM, Webster D, Foxwell BM (2006) Bruton's tyrosine kinase is required for TLR2 and TLR4-induced TNF, but not IL-6, production. J Immunol 176(6):3635–3641

Hsieh FH, Lam BK, Penrose JF, Austen KF, Boyce JA (2001) T helper cell type 2 cytokines coordinately regulate immunoglobulin E-dependent cysteinyl leukotriene production by human cord blood-derived mast cells: profound induction of leukotriene $C_{(4)}$ synthase expression by interleukin 4. J Exp Med 193(1):123–133

Hughes PJ, Holgate ST, Church MK (1984) Adenosine inhibits and potentiates IgE-dependent histamine release from human lung mast cells by an A2-purinoceptor mediated mechanism. Biochem Pharmacol 33(23):3847–3852

Isersky C, Taurog JD, Poy G, Metzger H (1978) Triggering of cultured neoplastic mast cells by antibodies to the receptor for IgE. J Immunol 121(2):549–558

Isersky C, Rivera J, Mims S, Triche TJ (1979) The fate of IgE bound to rat basophilic leukemia cells. J Immunol 122(5):1926–1936

Ishizaka T, Chang TH, Taggart M, Ishizaka K (1977) Histamine release from rat mast cells by antibodies against rat basophilic leukemia cell membrane. J Immunol 119(5):1589–1596

Ishizuka T, Kawasome H, Terada N, Takeda K, Gerwins P, Keller GM, Johnson GL, Gelfand EW (1998) Stem cell factor augments Fc ε RI-mediated TNF-α production and stimulates MAP kinases via a different pathway in MC/9 mast cells. J Immunol 161(7):3624–3630

Izawa K, Kitaura J, Yamanishi Y, Matsuoka T, Kaitani A, Sugiuchi M, Takahashi M, Maehara A, Enomoto Y, Oki T, Takai T, Kitamura T (2009) An activating and inhibitory signal from an inhibitory receptor LMIR3/CLM-1: LMIR3 augments lipopolysaccharide response through association with FcRγ in mast cells. J Immunol 183(2):925–936

Izawa K, Yamanishi Y, Maehara A, Takahashi M, Isobe M, Ito S, Kaitani A, Matsukawa T, Matsuoka T, Nakahara F, Oki T, Kiyonari H, Abe T, Okumura K, Kitamura T, Kitaura J (2012) The receptor LMIR3 negatively regulates mast cell activation and allergic responses by binding to extracellular ceramide. Immunity 37(5):827–839

Jackola DR, Pierson-Mullany LK, Liebeler CL, Blumenthal MN, Rosenberg A (2002) Variable binding affinities for allergen suggest a 'selective competition' among immunoglobulins in atopic and non-atopic humans. Mol Immunol 39(5–6):367–377

James LC, Roversi P, Tawfik DS (2003) Antibody multispecificity mediated by conformational diversity. Science 299(5611):1362–1367

Jefferies CA, Doyle S, Brunner C, Dunne A, Brint E, Wietek C, Walch E, Wirth T, O'Neill LA (2003) Bruton's tyrosine kinase is a Toll/interleukin-1 receptor domain-binding protein that participates in nuclear factor κB activation by Toll-like receptor 4. J Biol Chem 278 (28):26258–26264

Jimenez-Andrade GY, Ibarra-Sanchez A, Gonzalez D, Lamas M, Gonzalez-Espinosa C (2013) Immunoglobulin E induces VEGF production in mast cells and potentiates their protumorigenic actions through a Fyn kinase-dependent mechanism. J Hematol Oncol 6:56

Jolly PS, Bektas M, Olivera A, Gonzalez-Espinosa C, Proia RL, Rivera J, Milstien S, Spiegel S (2004) Transactivation of sphingosine-1-phosphate receptors by FcεRI triggering is required for normal mast cell degranulation and chemotaxis. J Exp Med 199(7):959–970

Kalesnikoff J, Huber M, Lam V, Damen JE, Zhang J, Siraganian RP, Krystal G (2001) Monomeric IgE stimulates signaling pathways in mast cells that lead to cytokine production and cell survival. Immunity 14(6):801–811

Kasakura K, Takahashi K, Aizawa T, Hosono A, Kaminogawa S (2009) A TLR2 ligand suppresses allergic inflammatory reactions by acting directly on mast cells. Int Arch Allergy Immunol 150(4):359–369

Kashiwakura J, Otani IM, Kawakami T (2011) Monomeric IgE and mast cell development, survival and function. Adv Exp Med Biol 716:29–46

Kashiwakura J, Okayama Y, Furue M, Kabashima K, Shimada S, Ra C, Siraganian RP, Kawakami Y, Kawakami T (2012) Most highly Cytokinergic IgEs have polyreactivity to autoantigens. Allergy Asthma Immunol Res 4(6):332–340

Katz HR, Vivier E, Castells MC, McCormick MJ, Chambers JM, Austen KF (1996) Mouse mast cell gp49B1 contains two immunoreceptor tyrosine-based inhibition motifs and suppresses mast cell activation when colligated with the high-affinity Fc receptor for IgE. Proc Natl Acad Sci USA 93(20):10809–10814

Kawakami T, Galli SJ (2002) Regulation of mast-cell and basophil function and survival by IgE. Nat Rev Immunol 2(10):773–786

Kawakami Y, Kitaura J, Hata D, Yao L, Kawakami T (1999) Functions of Bruton's tyrosine kinase in mast and B cells. J Leukoc Biol 65(3):286–290

Kennedy Norton S, Barnstein B, Brenzovich J, Bailey DP, Kashyap M, Speiran K, Ford J, Conrad D, Watowich S, Moralle MR, Kepley CL, Murray PJ, Ryan JJ (2008) IL-10 suppresses mast cell IgE receptor expression and signaling in vitro and in vivo. J Immunol 180(5):2848–2854

Kitaura J, Eto K, Kinoshita T, Kawakami Y, Leitges M, Lowell CA, Kawakami T (2005) Regulation of highly cytokinergic IgE-induced mast cell adhesion by Src, Syk, Tec, and protein kinase C family kinases. J Immunol 174(8):4495–4504

Kitaura J, Song J, Tsai M, Asai K, Maeda-Yamamoto M, Mocsai A, Kawakami Y, Liu FT, Lowell CA, Barisas BG, Galli SJ, Kawakami T (2003) Evidence that IgE molecules mediate a spectrum of effects on mast cell survival and activation via aggregation of the FcεRI. Proc Natl Acad Sci USA 100(22):12911–12916

Komiya A, Nagase H, Okugawa S, Ota Y, Suzukawa M, Kawakami A, Sekiya T, Matsushima K, Ohta K, Hirai K, Yamamoto K, Yamaguchi M (2006) Expression and function of toll-like receptors in human basophils. Int Arch Allergy Immunol 140(Suppl 1):23–27

Kuo CH, Morohoshi K, Aye CC, Garoon RB, Collins A, Ono SJ (2012) The role of TRB3 in mast cells sensitized with monomeric IgE. Exp Mol Pathol 93(3):408–415

Laffargue M, Calvez R, Finan P, Trifilieff A, Barbier M, Altruda F, Hirsch E, Wymann MP (2002) Phosphoinositide 3-kinase gamma is an essential amplifier of mast cell function. Immunity 16 (3):441–451

Lee JH, Kim JW, Kim do K, Kim HS, Park HJ, Park DK, Kim AR, Kim B, Beaven MA, Park KL, Kim YM, Choi WS (2011) The Src family kinase Fgr is critical for activation of mast cells and IgE-mediated anaphylaxis in mice. J Immunol 187(4):1807–1815

Liao BC, Hou RC, Wang JS, Jeng KC (2006) Enhancement of the release of inflammatory mediators by substance P in rat basophilic leukemia RBL-2H3 cells. J Biomed Sci 13(5):613–619

Lienard H, Bruhns P, Malbec O, Fridman WH, Daeron M (1999) Signal regulatory proteins negatively regulate immunoreceptor-dependent cell activation. J Biol Chem 274(45):32493–32499

Lin S, Cicala C, Scharenberg AM, Kinet JP (1996) The FcεRIβ subunit functions as an amplifier of FceRIg-mediated cell activation signals. Cell 85(7):985–995

Liu Y, Furuta K, Teshima R, Shirata N, Sugimoto Y, Ichikawa A, Tanaka S (2005) Critical role of protein kinase C betaII in activation of mast cells by monomeric IgE. J Biol Chem 280 (47):38976–38981

Madrenas J, Wange RL, Wang JL, Isakov N, Samelson LE, Germain RN (1995) Zeta phosphorylation without ZAP-70 activation induced by TCR antagonists or partial agonists. Science 267(5197):515–518

Marquardt DL, Parker CW, Sullivan TJ (1978) Potentiation of mast cell mediator release by adenosine. J Immunol 120(3):871–878

Matsuda K, Piliponsky AM, Iikura M, Nakae S, Wang EW, Dutta SM, Kawakami T, Tsai M, Galli SJ (2005) Monomeric IgE enhances human mast cell chemokine production: IL-4 augments and dexamethasone suppresses the response. J Allergy Clin Immunol 116(6):1357–1363

Matsushima H, Yamada N, Matsue H, Shimada S (2004) TLR3-, TLR7-, and TLR9-mediated production of proinflammatory cytokines and chemokines from murine connective tissue type skin-derived mast cells but not from bone marrow-derived mast cells. J Immunol 173(1):531–541

McCary C, Tancowny BP, Catalli A, Grammer LC, Harris KE, Schleimer RP, Kulka M (2010) Substance P downregulates expression of the high affinity IgE receptor (FcεRI) by human mast cells. J Neuroimmunol 220(1–2):17–24

McKeithan TW (1995) Kinetic proofreading in T-cell receptor signal transduction. Proc Natl Acad Sci USA 92(11):5042–5046

Melendez AJ, Harnett MM, Pushparaj PN, Wong WS, Tay HK, McSharry CP, Harnett W (2007) Inhibition of Fc ε RI-mediated mast cell responses by ES-62, a product of parasitic filarial nematodes. Nat Med 13(11):1375–1381

Metzger H (1977) The cellular receptor for IgE. In: Cuatrecasas P, Greaves MF (eds) Receptors and recognition. Chapman and Hall, London, pp 75–102

Metzger H (1978) The IgE-mast cell system as a paradigm for the study of antibody mechanisms. Immunol Rev 41:186–199

Metzger H (1983) The receptor on mast cells and related cells with high affinity for IgE. Contemp Top Mol Immunol 9:115–145

Metzger H (1992) Transmembrane signaling: the joy of aggregation. J Immunol 149(5):1477–1487

Meyaard L (2008) The inhibitory collagen receptor LAIR-1 (CD305). J Leukoc Biol 83(4):799–803

Miyazaki D, Nakamura T, Toda M, Cheung-Chau KW, Richardson RM, Ono SJ (2005) Macrophage inflammatory protein-1α as a costimulatory signal for mast cell-mediated immediate hypersensitivity reactions. J Clin Invest 115(2):434–442

Nadler MJ, Kinet JP (2002) Uncovering new complexities in mast cell signaling. Nat Immunol 3(8):707–708

Nakae S, Iikura M, Suto H, Akiba H, Umetsu DT, Dekruyff RH, Saito H, Galli SJ (2007) TIM-1 and TIM-3 enhancement of Th2 cytokine production by mast cells. Blood 110(7):2565–2568

Nguyen M, Solle M, Audoly LP, Tilley SL, Stock JL, McNeish JD, Coffman TM, Dombrowicz D, Koller BH (2002) Receptors and signaling mechanisms required for prostaglandin E2-mediated regulation of mast cell degranulation and IL-6 production. J Immunol 169(8):4586–4593

Nishimoto H, Lee SW, Hong H, Potter KG, Maeda-Yamamoto M, Kinoshita T, Kawakami Y, Mittler RS, Kwon BS, Ware CF, Croft M, Kawakami T (2005) Costimulation of mast cells by 4-1BB, a member of the tumor necrosis factor receptor superfamily, with the high-affinity IgE receptor. Blood 106(13):4241–4248

Novak N, Bieber T, Peng WM (2010) The immunoglobulin E-Toll-like receptor network. Int Arch Allergy Immunol 151(1):1–7

Ochi H, De Jesus NH, Hsieh FH, Austen KF, Boyce JA (2000) IL-4 and-5 prime human mast cells for different profiles of IgE-dependent cytokine production. Proc Natl Acad Sci USA 97(19):10509–10513

Okumura S, Kashiwakura J, Tomita H, Matsumoto K, Nakajima T, Saito H, Okayama Y (2003) Identification of specific gene expression profiles in human mast cells mediated by Toll-like receptor 4 and FcεRI. Blood 102(7):2547–2554

Olivera A (2008) Unraveling the complexities of sphingosine-1-phosphate function: the mast cell model. Prostaglandins Other Lipid Mediat 86(1–4):1–11

Olivera A, Mizugishi K, Tikhonova A, Ciaccia L, Odom S, Proia RL, Rivera J (2007) The sphingosine kinase-sphingosine-1-phosphate axis is a determinant of mast cell function and anaphylaxis. Immunity 26(3):287–297

Olivera A, Rivera J (2011) An emerging role for the lipid mediator sphingosine-1-phosphate in mast cell effector function and allergic disease. Adv Exp Med Biol 716:123–142

On M, Billingsley JM, Jouvin MH, Kinet JP (2004) Molecular dissection of the FcRβ signaling amplifier. J Biol Chem 279(44):45782–45790

Ortega Soto E, Pecht I (1988) A monoclonal antibody that inhibits secretion from rat basophilic leukemia cells and binds to a novel membrane component. J Immunol 141(12):4324–4332

Oskeritzian CA, Price MM, Hait NC, Kapitonov D, Falanga YT, Morales JK, Ryan JJ, Milstien S, Spiegel S (2010) Essential roles of sphingosine-1-phosphate receptor 2 in human mast cell activation, anaphylaxis, and pulmonary edema. J Exp Med 207(3):465–474

Paar JM, Harris NT, Holowka D, Baird B (2002) Bivalent ligands with rigid double-stranded DNA spacers reveal structural constraints on signaling by Fc ε RI. J Immunol 169(2):856–864

Paolini R, Serra A, Kinet JP (1996) Persistence of tyrosine-phosphorylated FcεRI in deactivated cells. J Biol Chem 271(27):15987–15992

Parravicini V, Gadina M, Kovarova M, Odom S, Gonzalez-Espinosa C, Furumoto Y, Saitoh S, Samelson LE, O'Shea JJ, Rivera J (2002) Fyn kinase initiates complementary signals required for IgE-dependent mast cell degranulation. Nat Immunol 3(8):741–748

Pasquier B, Launay P, Kanamaru Y, Moura IC, Pfirsch S, Ruffie C, Henin D, Benhamou M, Pretolani M, Blank U, Monteiro RC (2005) Identification of FcαRI as an inhibitory receptor that controls inflammation: dual role of FcRγ ITAM. Immunity 22(1):31–42

Pietrzak A, Wierzbicki M, Wiktorska M, Brzezinska-Blaszczyk E (2011) Surface TLR2 and TLR4 expression on mature rat mast cells can be affected by some bacterial components and proinflammatory cytokines. Mediators Inflamm 2011:427473

Pribluda VS, Pribluda C, Metzger H (1994) Transphosphorylation as the mechanism by which the high-affinity receptor for IgE is phosphorylated upon aggregation. Proc Natl Acad Sci USA 91 (23):11246–11550

Price KS, Friend DS, Mellor EA, De Jesus N, Watts GF, Boyce JA (2002) CC chemokine receptor 3 mobilizes to the surface of human mast cells and potentiates immunoglobulin E-dependent generation of interleukin 13. Am J Respir Cell Mol Biol 28(4):420–427

Price LS, Norman JC, Ridley AJ, Koffer A (1995) The small GTPases Rac and Rho as regulators of secretion in mast cells. Curr Biol 5(1):68–73

Pyenta PS, Schwille P, Webb WW, Holowka D, Baird B (2003) Lateral diffusion of membrane lipid-anchored probes before and after aggregation of cell surface IgE-receptors. J Phys Chem A 107:8310–8318

Qiao H, Andrade MV, Lisboa FA, Morgan K, Beaven MA (2006) FcεR1 and toll-like receptors mediate synergistic signals to markedly augment production of inflammatory cytokines in murine mast cells. Blood 107(2):610–618

Ramkumar V, Stiles GL, Beaven MA, Ali H (1993) The A3 adenosine receptor is the unique adenosine receptor which facilitates release of allergic mediators in mast cells. J Biol Chem 268(23):16887–16890

Rivera J, Fierro NA, Olivera A, Suzuki R (2008) New insights on mast cell activation via the high affinity receptor for IgE. Adv Immunol 98:85–120

Ryzhov S, Goldstein AE, Biaggioni I, Feoktistov I (2006) Cross-talk between $G_{(s)}$- and $G_{(q)}$-coupled pathways in regulation of interleukin-4 by $A_{(2B)}$ adenosine receptors in human mast cells. Mol Pharmacol 70(2):727–735

Saitoh S, Arudchandran R, Manetz TS, Zhang W, Sommers CL, Love PE, Rivera J, Samelson LE (2000) LAT is essential for FcεRI-mediated mast cell activation. Immunity 12(5):525–535

Schulman ES, Glaum MC, Post T, Wang Y, Raible DG, Mohanty J, Butterfield JH, Pelleg A (1999) ATP modulates anti-IgE-induced release of histamine from human lung mast cells. Am J Respir Cell Mol Biol 20(3):530–537

Selander C, Engblom C, Nilsson G, Scheynius A, Andersson CL (2009) TLR2/MyD88-dependent and -independent activation of mast cell IgE responses by the skin commensal yeast Malassezia sympodialis. J Immunol 182(7):4208–4216

Serra-Pages M, Olivera A, Torres R, Picado C, de Mora F, Rivera J (2012) E-prostanoid 2 receptors dampen mast cell degranulation via cAMP/PKA-mediated suppression of IgE-dependent signaling. J Leukoc Biol 92(6):1155–1165

Sibilano R, Frossi B, Suzuki R, D'Inca F, Gri G, Piconese S, Colombo MP, Rivera J, Pucillo CE (2012) Modulation of FcεRI-dependent mast cell response by OX40L via Fyn, PI3 K, and RhoA. J Allergy Clin Immunol 130(3):751–760e2

Sil D, Lee JB, Luo D, Holowka D, Baird B (2007) Trivalent ligands with rigid DNA spacers reveal structural requirements for IgE receptor signaling in RBL mast cells. ACS Chem Biol 2 (10):674–684

Sloan-Lancaster J, Shaw AS, Rothbard JB, Allen PM (1994) Partial T cell signaling: altered phospho-zeta and lack of zap70 recruitment in APL-induced T cell anergy. Cell 79(5):913–922

Sonnenblick A, Levy C, Razin E (2005) Immunological trigger of mast cells by monomeric IgE: effect on microphthalmia transcription factor, STAT3 network of interactions. J Immunol 175 (3):1450–1455

Suzuki R, Leach S, Dema B, Rivera J (2013) Characterization of a phospho-specific antibody to the Fcε receptor γ chain, reveals differences in the regulation of Syk and Akt phosphorylation. Antibodies 2:321–337

Suzuki R, Leach S, Liu W, Ralston E, Scheffel J, Zhang W, Lowell CA, Rivera J (2014) Molecular editing of cellular responses by the high-affinity receptor for IgE. Science 343(6174):1021–1025

Tashiro M, Kawakami Y, Abe R, Han W, Hata D, Sugie K, Yao L, Kawakami T (1997) Increased secretion of TNF-alpha by costimulation of mast cells via CD28 and Fc ε RI. J Immunol 158 (5):2382–2389

Thomas JL, Holowka D, Baird B, Webb WW (1994) Large-scale co-aggregation of fluorescent lipid probes with cell surface proteins. J Cell Biol 125(4):795–802

Tkaczyk C, Horejsi V, Iwaki S, Draber P, Samelson LE, Satterthwaite AB, Nahm DH, Metcalfe DD, Gilfillan AM (2004) NTAL phosphorylation is a pivotal link between the signaling cascades leading to human mast cell degranulation following Kit activation and Fc ε RI aggregation. Blood 104(1):207–214

Toda M, Dawson M, Nakamura T, Munro PM, Richardson RM, Bailly M, Ono SJ (2004) Impact of engagement of FcεRI and CC chemokine receptor 1 on mast cell activation and motility. J Biol Chem 279(46):48443–48448

Tolar P, Hanna J, Krueger PD, Pierce SK (2009) The constant region of the membrane immunoglobulin mediates B cell-receptor clustering and signaling in response to membrane antigens. Immunity 30(1):44–55

Torigoe C, Inman JK, Metzger H (1998) An unusual mechanism for ligand antagonism. Science 281(5376):568–572

Toru H, Pawankar R, Ra C, Yata J, Nakahata T (1998) Human mast cells produce IL-13 by high-affinity IgE receptor cross-linking: enhanced IL-13 production by IL-4-primed human mast cells. J Allergy Clin Immunol 102(3):491–502

Uehara T, Blery M, Kang DW, Chen CC, Ho LH, Gartland GL, Liu FT, Vivier E, Cooper MD, Kubagawa H (2001) Inhibition of IgE-mediated mast cell activation by the paired Ig-like receptor PIR-B. J Clin Invest 108(7):1041–1050

van der Merwe PA, Davis SJ (2003) Molecular interactions mediating T cell antigen recognition. Annu Rev Immunol 21:659–684

Varadaradjalou S, Feger F, Thieblemont N, Hamouda NB, Pleau JM, Dy M, Arock M (2003) Toll-like receptor 2 (TLR2) and TLR4 differentially activate human mast cells. Eur J Immunol 33 (4):899–906

Vonakis BM, Chen H, Haleem-Smith H, Metzger H (1997) The unique domain as the site on Lyn kinase for its constitutive association with the high affinity receptor for IgE. J Biol Chem 272 (38):24072–24080

Wang XS, Lau HY (2006) Prostaglandin E potentiates the immunologically stimulated histamine release from human peripheral blood-derived mast cells through EP1/EP3 receptors. Allergy 61 (4):503–506

Weetall M, Holowka D, Baird B (1993) Heterologous desensitization of the high affinity receptor for IgE (FcεRI) on RBL cells. J Immunol 150(9):4072–4083

Wong MX, Roberts D, Bartley PA, Jackson DE (2002) Absence of platelet endothelial cell adhesion molecule-1 (CD31) leads to increased severity of local and systemic IgE-mediated anaphylaxis and modulation of mast cell activation. J Immunol 168(12):6455–6462

Woolhiser MR, Brockow K, Metcalfe DD (2004) Activation of human mast cells by aggregated IgG through FcγRI: additive effects of C3a. Clin Immunol 110(2):172–180

Woolhiser MR, Okayama Y, Gilfillan AM, Metcalfe DD (2001) IgG-dependent activation of human mast cells following up-regulation of FcγRI by IFN-γ. Eur J Immunol 31(11):3298–3307

Yamaguchi M, Lantz CS, Oettgen HC, Katona IM, Fleming T, Miyajima I, Kinet JP, Galli SJ (1997) IgE enhances mouse mast cell FcεRI expression in vitro and in vivo: evidence for a novel amplification mechanism in IgE-dependent reactions. J Exp Med 185(4):663–672

Yamasaki S, Ishikawa E, Kohno M, Saito T (2004) The quantity and duration of FcRgamma signals determine mast cell degranulation and survival. Blood 103(8):3093–3101

Yamashita T, Suzuki R, Backlund PS, Yamashita Y, Yergey AL, Rivera J (2008) Differential dephosphorylation of the FcRγ immunoreceptor tyrosine-based activation motif tyrosines with dissimilar potential for activating Syk. J Biol Chem 283(42):28584–28594

Yang Z, Yan WX, Cai H, Tedla N, Armishaw C, Di Girolamo N, Wang HW, Hampartzoumian T, Simpson JL, Gibson PG, Hunt J, Hart P, Hughes JM, Perry MA, Alewood PF, Geczy CL (2007) S100A12 provokes mast cell activation: a potential amplification pathway in asthma and innate immunity. J Allergy Clin Immunol 119(1):106–114

Yokoi H, Choi OH, Hubbard W, Lee HS, Canning BJ, Lee HH, Ryu SD, von Gunten S, Bickel CA, Hudson SA, Macglashan DW, Jr, Bochner BS (2008) Inhibition of FcεRI-dependent mediator release and calcium flux from human mast cells by sialic acid-binding immunoglobulin-like lectin 8 engagement. J Allergy Clin Immunol 121(2):499–505e1

Yoshioka M, Fukuishi N, Iriguchi S, Ohsaki K, Yamanobe H, Inukai A, Kurihara D, Imajo N, Yasui Y, Matsui N, Tsujita T, Ishii A, Seya T, Takahama M, Akagi M (2007) Lipoteichoic acid downregulates FcεRI expression on human mast cells through Toll-like receptor 2. J Allergy Clin Immunol 120(2):452–461

Yotsumoto K, Okoshi Y, Shibuya K, Yamazaki S, Tahara-Hanaoka S, Honda S, Osawa M, Kuroiwa A, Matsuda Y, Tenen DG, Iwama A, Nakauchi H, Shibuya A (2003) Paired activating and inhibitory immunoglobulin-like receptors, MAIR-I and MAIR-II, regulate mast cell and macrophage activation. J Exp Med 198(2):223–233

Young RM, Holowka D, Baird B (2003) A lipid raft environment enhances Lyn kinase activity by protecting the active site tyrosine from dephosphorylation. J Biol Chem 278(23):20746–20752

Young RM, Zheng X, Holowka D, Baird B (2005) Reconstitution of regulated phosphorylation of FcεRI by a lipid raft-excluded protein-tyrosine phosphatase. J Biol Chem 280(2):1230–1235

Zhang S, Cherwinski H, Sedgwick JD, Phillips JH (2004) Molecular mechanisms of CD200 inhibition of mast cell activation. J Immunol 173(11):6786–6793

Zorn CN, Keck S, Hendriks RW, Leitges M, Freudenberg MA, Huber M (2009) Bruton's tyrosine kinase is dispensable for the toll-like receptor-mediated activation of mast cells. Cell Signal 21 (1):79–86

Helminth-Induced IgE and Protection Against Allergic Disorders

Firdaus Hamid, Abena S. Amoah, Ronald van Ree and Maria Yazdanbakhsh

Abstract The immune response against helminths and allergens is generally characterized by high levels of IgE and increased numbers of Th2 cells, eosinophils, and mast cells, yet the clinical outcome with respect to immediate hypersensitivity and inflammation is clearly not the same. High levels of IgE are seen to allergens during helminth infections; however, these IgE responses do not translate into allergy symptoms. This chapter summarizes the evidence of the association between helminth infections and allergic disorders. It discusses how helminth infection can lead to IgE cross-reactivity with allergens and how this IgE has poor biological activity. This information is important for developing new diagnostic methods and treatments for allergic disorders in low-to-middle-income countries.

F. Hamid
Department of Microbiology, Faculty of Medicine, Hasanuddin University, Makassar, Indonesia
e-mail: firdaus.hamid@gmail.com; f.hamid@lumc.nl

F. Hamid · A.S. Amoah · M. Yazdanbakhsh (✉)
Department of Parasitology, Leiden University Medical Center, Albinusdreef 2, 2333, ZA Leiden, The Netherlands
e-mail: m.yazdanbakhsh@lumc.nl

A.S. Amoah
e-mail: chardona@gmail.com; a.s.amoah@lumc.nl

A.S. Amoah
Department of Parasitology, Noguchi Memorial Institute for Medical Research, Accra, Ghana

R. van Ree
Department of Experimental Immunology and Department of Otorhinolaryngology, Academic Medical Center, Amsterdam University, Amsterdam, The Netherlands
e-mail: r.vanree@amc.uva.nl

© Springer International Publishing Switzerland 2015
J.J. Lafaille and M.A. Curotto de Lafaille (eds.), *IgE Antibodies: Generation and Function*, Current Topics in Microbiology and Immunology 388,
DOI 10.1007/978-3-319-13725-4_5

Contents

1 Introduction 92
2 Helminth Infections and Allergic Disorders 93
   2.1 Mechanisms Behind the Association Between Helminths and Allergies 97
3 Cross-Reactivity Between Allergen and Helminths 98
   3.1 Cross-Reactive Carbohydrate Determinants (CCDs) and Helminths 98
   3.2 Peptide Cross-Reactivity and Helminths 100
4 Component-Resolved Diagnosis (CRD) in Allergy Diagnosis 102
   4.1 Microarray 103
5 Future Directions 103
References 104

1 Introduction

Over the past few decades, the prevalence of allergic disease has been on the rise in both developed and developing countries (Pawankar et al. 2011). In addition, many epidemiological studies have shown a higher prevalence of allergic disorders in subjects living in urban environments compared to rural areas, particularly in developing countries (Hamid et al. 2013; Perzanowski et al. 2002).

Such observations indicate that environmental factors, along with underlying genetics, play a key role in the development of allergic disease. Environmental influences can include exposures to microbes, parasites, and lifestyle factors (Burke et al. 2003; von Mutius 2002). Of particular interest have been infections with parasitic helminths that are highly prevalent in tropical regions of the developing world. It is estimated that a quarter of the World's population is chronically infected by helminths such as *Ascaris lumbricoides* (roundworm), *Trichuris trichiura* (whipworm), *Necator americanus* or *Ancylostoma duodenale* (hookworms), schistosomes, and filarial worms (Bethony et al. 2006).

Despite the close parallels between immune responses that characterize helminth infections and allergic diseases, namely increased levels of immunoglobulin (Ig)-E, eosinophils, and mast cells along with T cells that preferentially secrete T helper type 2 (Th2) cytokines, the clinical outcome with respect to immediate hypersensitivity and inflammation is not the same (Yazdanbakhsh et al. 2001). Moreover, there is little geographical overlap worldwide between helminth infections and allergies. In fact, several studies have reported a negative association between the presence of helminth infections and allergic disorders (Feary et al. 2011). However, the relationship between allergic disorders and helminth infections does not show consistent results since there are also studies that show that helminth infections either have no effect or are associated with increased atopic disorders (Obihara et al. 2006; Palmer et al. 2002).

High levels of Th2 responses can lead to allergic diseases, and to prevent this, either a shift to Th1 or an increased anti-inflammatory response would be needed (Carvalho et al. 2006; Cooper 2009). A number of immune mechanisms have been proposed to account for the negative association between helminths and allergies. The observations that chronic helminth infections are associated with higher suppressive responses, such as interleukin-10 (IL-10) and regulatory T cells, led to the proposal that a strong regulatory network induced by helminths might prevent the downstream effector phase of Th2 responses, preventing excessive inflammation (Satoguina et al. 2008; Wilson et al. 2005).

Another mechanism that might explain the inverse association between helminth infections and allergies could involve helminth-induced IgE. Early studies have suggested that polyclonal IgE that is stimulated by helminth infections might compete with allergen-specific IgE and therefore block degranulation of basophils and mast cells. However, a number of studies have refuted this IgE-blocking hypothesis. The idea that IgG4 antibodies associated with helminth infections might bind allergen and quench it from binding to IgE was put forward by Aalberse et al. (2009). An alternative idea that helminth infections may be associated with increased levels of allergen-specific IgE that are functionally poor and therefore cannot lead to basophil or mast cell degranulation (Yazdanbakhsh et al. 2002) is gaining support from several recent studies that indicate cross-reactive IgE might be associated with poor biological activity (Amoah et al. 2013b; Larson et al. 2012a).

Over the years, there has been increasing number of publications covering (a) the relationship between helminths and IgE, which started in 1969; (b) IgE and the issue of cross-reactivity; and (c) component-resolved diagnosis (CRD) (Fig. 1). In this chapter, we will focus on (1) the epidemiological studies of the relationship between helminth infections and allergic disorders and (2) the characterization of helminth-induced IgE and the possible application of new technologies in allergy diagnostics in low-to-middle-income countries.

2 Helminth Infections and Allergic Disorders

The prevalence of allergic diseases has been increasing mainly in the developed world where long ago changes have been seen in lifestyle and the environment characterized by increasing sanitation, hygienic measures, and urbanization. In less developed countries, the prevalence of allergic disorders is relatively low, but the allergic march is starting in these geographical areas due to ongoing dramatic changes in lifestyle and the environment. Although not as dramatic as in Western countries, an increase in allergic disorders has been reported in developing countries with the tendency for prevalences in urban centers to approach those seen in affluent countries.

In developed countries, the presence of IgE antibodies to allergens increases the risk of allergic disorders. These antibodies, which bind to high-affinity IgE receptors present on basophils and mast cells, can be cross-linked by allergens, an event that leads to mast cell degranulation and histamine release. The mast cell degranulation leads to inflammation and in target organs such as the airways can result in symptoms which can be recognized as an asthmatic attack. One of the diagnostic

Number of publications on topics covered by this chapter

— Helminth and IgE
— IgE cross-reactivity
— component-resolved diagnosis

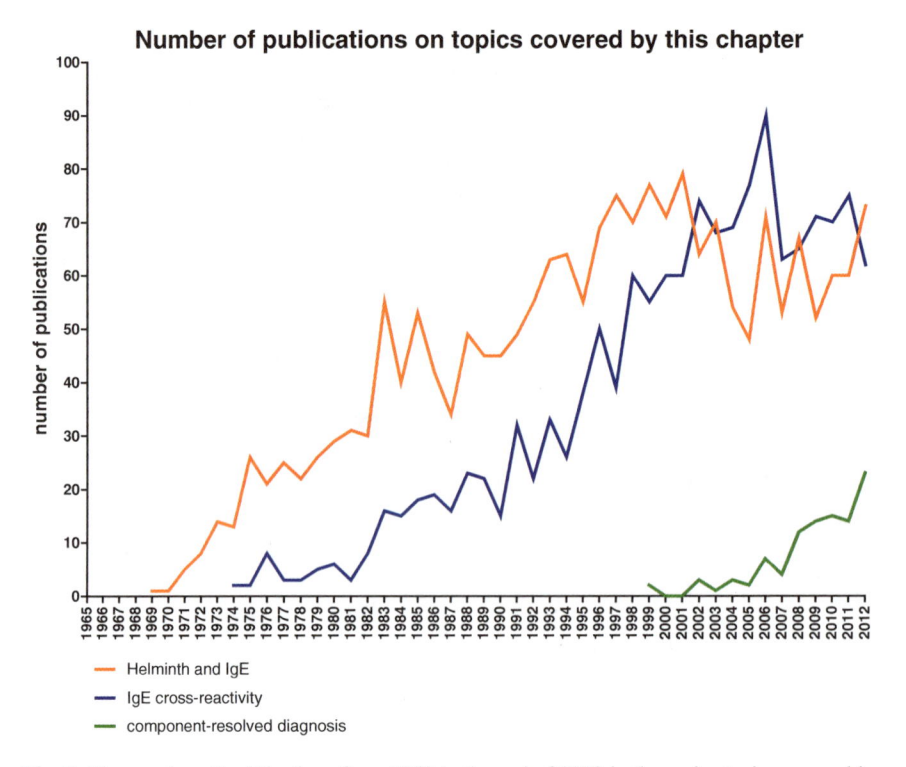

Fig. 1 The number of publications from 1969 to the end of 2012 in the major topics covered by this chapter. Specific search terms used in PubMed were "Helminth and IgE" (*orange*), "IgE cross-reactivity" (*blue*), and "component-resolved diagnosis" (*green*)

methods for allergy is to perform a skin prick test (SPT). This is based on applying allergens to the skin, pricking the skin with a lancet and assessing whether a reaction develops to the allergen within 15 min. If a wheal and flare easily visible in lighter skin reaction is seen, which is the result of mast cell degranulation, then it is concluded that person being tested is sensitized to the allergen applied. In developed countries, the SPT is often interchangeable with the measurement of IgE antibodies to allergens by in vitro tests such as the ImmunoCAP assay. This means that allergen-specific IgE measured in serum can be associated with SPT reactivity and potentially with allergic symptoms. However, the relationship between IgE, SPT, and symptoms of allergy can be different in different geographical areas. For example, the proportion of SPT reactivity with clinical allergic asthma appears to be much smaller in some rural areas compared to urban centers in Europe (Priftanji et al. 2001) or in many developing nations compared to developed ones (Cooper et al. 2003a; Dagoye et al. 2003). Helminth infections have some interesting effects on the relationship between IgE, SPT, and clinical symptoms.

Several epidemiological studies have shown that helminth infections can be negatively associated with allergic outcomes (Feary et al. 2011). Some studies have shown that having schistosome or filarial infections decreases the risk of SPT positivity (Supali et al. 2010; van den Biggelaar et al. 2001). Similarly, studies in Ecuadorian and Vietnamese children have demonstrated that having soil-transmitted helminth infections decreases the risk of SPT positivity (Cooper et al. 2004; Flohr et al. 2006). In addition, Cooper et al. (2003b) found that the presence of serological markers of chronic infections (elevated levels of total serum IgE or anti-*A. lumbricoides* IgG4) was independently negatively associated with allergen SPT reactivity. Araujo et al. (2000) reported a strong and inverse association between skin responses to allergens and infection with *Schistosoma mansoni*, among persons living in an area endemic for this helminth (Araujo et al. 2000). Similarly, a study in Gabonese children observed that the risk of a positive SPT was reduced by 72 % if a child was infected with *S. haematobium*, a blood-dwelling helminth (van den Biggelaar et al. 2001). Moreover, an investigation conducted among 1,385 urban and rural Ghanaian children aged 5–16 years showed a strong negative association between schistosome infection and SPT reactivity to mite but not with reported wheeze or asthma (Obeng et al. 2014). However, with regard to clinical symptoms of asthma, a case–control study in urban and rural Ethiopians aged 16–60 years determined that active hookworm infection reduced the risk of reported wheeze (Scrivener et al. 2001).

Although the majority of studies have shown a negative association between helminth infection and SPT, there are studies that show that helminth may increase the risk of asthma and atopic disorder. Obihara et al. (2006) found *Ascaris*-specific IgE may be a risk factor for atopic disease in populations exposed to mild *A. lumbricoides* infection (Obihara et al. 2006). In line with this, a study in China, in an area with a low burden of *A. lumbricoides*, demonstrated a positive association between helminth and allergen skin test reactivity as well as asthma risk (Palmer et al. 2002). Finally, there are also studies showing no significant association. A study in Brazil among patients aged 12–30 years with asthma or rhinitis living in an urban area endemic for geohelminth showed that individuals infected with a low parasite burden of *A. lumbricoides* did not differ on the frequency of positive SPT to dust mites from those *A. lumbricoides* negatives living in the same area (Ponte et al. 2006). A birth cohort of children from Ethiopia which investigated the effect of geohelminths on allergic symptoms, at 3 years of age, found that there was no association between helminth infections and wheeze nor with eczema (Amberbir et al. 2011).

Considering that cross-sectional studies can only show relationships, it is important to prove that helminths, and no other confounders, are responsible for any association seen with allergies. For this, interventional studies such as use of anthelmintics to remove helminths or intentional infection with helminths are needed. With respect to anthelmintic treatment, a study of soil-transmitted helminth-infected subjects demonstrated that regular anthelmintic treatment resulted in significant increase in skin test reactivity as well as serum level of IgE to aeroallergens (Lynch et al. 1993). In line with this, anthelmintic treatment of Gabonese children chronically infected with *S. haematobium* and soil-transmitted helminths resulted in increased SPT reactivity (van den Biggelaar et al. 2000). Three large interventional studies conducted in Ecuador, Vietnam, and Indonesia have shown different results.

In Ecuador, treatment with albendazole every 2 months, for 1 year, did not affect SPT nor clinical symptoms of allergy (Cooper et al. 2006), but in Vietnam, three-monthly treatment with albendazole resulted in a significant increase in SPT positivity but not in allergy symptoms (Flohr et al. 2010). The third study in Indonesia revealed that intensive community treatment of 3-monthly albendazole for 21 months over 2 years was not associated with increased risk of SPT to any allergen, but post hoc analysis showed that SPT to cockroach allergen was increased in the albendazole arm compared to placebo (Wiria et al. 2013). However, in agreement with the studies in Ecuador and Vietnam, there was no effect on reported clinical symptoms of allergy (Wiria et al. 2013). However, it has to be noted that one study in Venezuela showed that anthelmintic treatment resulted in improvement in all clinical indicators of asthma (Lynch et al. 1997). On the other hand, when considering anthelmintic treatment given during pregnancy, a large randomized, double-blind, placebo-controlled trial carried out in Uganda found that treatment of pregnant women with albendazole (compared with placebo) was strongly linked to an increased risk of doctor-diagnosed infantile eczema in their infants (Mpairwe et al. 2011).

Another approach to investigate the effect of helminths on allergies would be to use data from studies where humans have been experimentally infected with helminths. Recent years have seen an increasing number of such studies where volunteers were infected with L3 larvae of hookworm *N. americanus* as well *T. suis* eggs (Jouvin and Kinet 2012; Wright and Bickle 2005). From these studies, it is possible to delineate whether helminth infections are associated with increased allergies or whether they can suppress allergic symptoms. Helminth infections lead to expansion of Th2 responses and upregulation of IgE as well as eosinophils (Wright and Bickle 2005). This expansion of Th2 responses was not associated with an increase in allergic symptoms even though the life cycle of the hookworm parasite involves lung passage. In two safety trials, small numbers of patients with rhinoconjunctivitis or asthma were treated with helminths. Ten L3 infective larvae of *N. americanus* were inoculated into allergic patients in UK who were followed up for 16 weeks and showed no worsening or improvement of their symptoms (Feary et al. 2010). Regarding *T. suis* ova therapy, a double-blind placebo-controlled trial conducted in Danish adults assessed the efficacy of *T. suis* ova therapy for the treatment of grass pollen-induced allergic rhinitis and found no therapeutic effect of this particular therapy (Bager et al. 2010).

Taken together, most studies, but not all, show that helminth infections are associated with decreased SPT, but there does not seem to be a strong effect on clinical symptoms with the possible exception of a beneficial effect on infantile eczema. Both for anthelmintic treatment studies and experimental helminth infections, it is possible that different helminths with their varying life cycles and locations in tissues would lead to different effects on allergic outcomes. Moreover, it should be noted that chronicity of infection as well as worm burden might be important parameters to take into account when studying the association between helminths and allergies. Chronic infections as well as higher worm burdens might have stronger regulatory effect on allergies than acute or light infections (Smits et al. 2007). Finally, attention should be paid to the methods used to assess clinical symptoms of allergy as these could be a source of variability.

2.1 Mechanisms Behind the Association Between Helminths and Allergies

Given that most studies seem to show a negative association between chronic helminth infections and SPT reactivity, researchers have been looking for the mechanisms that could explain this. Although the mechanisms associated with this inverse relationship are not fully understood, it has been suggested that strong immune regulatory networks might be involved. This means that high levels of suppressive cytokines such as IL-10 and transforming growth factor-beta (TGF-β) as well as regulatory T and B cells (Smits et al. 2010), which seem to expand during chronic infections with helminth parasites, might downregulate allergic responses. The question as to where in the allergy cascade they exert their downregulatory activity is still unanswered. It is possible that regulatory responses affect Th2 and thereby IgE. However, immune regulation could also downregulate the effector phase of an allergic response which involves inflammation induced by mast cell degranulation (Larson et al. 2012b). This notion is supported by reports showing that IL-10 could inhibit basophil degranulation (Royer et al. 2001) and by the negative association between IL-10 and SPT (Macaubas et al. 1999; van den Biggelaar et al. 2000). Regarding antibodies, in the 1970s, the idea that polyclonal stimulation of IgE-producing plasma cells with many different specificities would compete with allergen-specific IgE for binding to high-affinity IgE receptors on mast cells was first proposed (Lancet editorial 1976; Godfrey 1975). This competition would reduce the chance of an allergen-dependent mast cell degranulation and therefore explain the absence of strong allergic responses in helminth-infected subjects (Lynch et al. 1998). However, a study by Mitre et al. (2005) showed that the high ration of polyclonal IgE to allergen-specific IgE did not inhibit basophil degranulation. Moreover, it has been shown that increases in IgE result in the upregulation of IgE receptors and that anti-IgE treatment is often accompanied by downregulation of IgE receptors which argues against the ability of high total IgE to compete out specific IgE (MacGlashan et al. 1997). Another hypothesis put forward is that high levels of IgG4 produced during parasitic infections could act as "blocking antibodies," since IgG4 is a Th2-dependent isotype not associated with clinical allergy (Aalberse et al. 2009; Arruda and Santos 2005). However, here, we will consider, yet another hypothesis, that helminths are associated with an IgE response that is cross-reactive with a low biological activity and therefore associated with less SPT reactivity to allergens and no strong increase in allergic symptoms in a Th2-skewed population.

3 Cross-Reactivity Between Allergen and Helminths

In general, cross-reactivity reflects the phylogenetic relationship between organisms that results in a high degree of homology in the primary structure of proteins and potentially in cross-reactivity (Aalberse et al. 2001). Cross-reactivity occurs when antibodies elicited to one epitope also recognize similar epitopes in other homologous molecules (Acevedo and Caraballo 2011). In allergy, the allergen that is supposed to induce the original allergic responses is named the primary sensitizer, and the others are considered cross-reactive allergens (Acevedo and Caraballo 2011). Two types of IgE cross-reactivities have been described: one cross-reactivity due to sugar moieties (glycans on glycoproteins) known as cross-reactive carbohydrate determinants (CCDs) and the other cross-reactivity due to proteins (Aalberse et al. 2001).

3.1 Cross-Reactive Carbohydrate Determinants (CCDs) and Helminths

The asparagine-linked carbohydrate components of plant and insect glycoproteins are highly cross-reactive and are known as CCDs (Altmann 2007). Two typical non-mammalian substitutions to N-glycans of plant glycoproteins are an $\alpha(1,3)$-linked fucose on the proximal N-acetyl glucosamine and a $\beta(1,2)$-linked xylose on the core mannose (van Ree et al. 2000).

These epitopes are also found in helminth parasites. The existence of IgE antibodies directed to CCDs was first reported by Aalberse et al. (1981) in the early 1980s. This study demonstrated that serum IgE from European pollen or venom-allergic patients cross-reacted with extracts from various allergenic foods. However, treating the extracts with periodate, which destroys the carbohydrate structures, abolished the reactions, indicating the involvement of carbohydrates in this cross-reactivity (Aalberse et al. 1981). Another investigation by the same group observed elevated levels of IgE against peanut extract among grass pollen-sensitized European patients without peanut SPT reactivity or clinical symptoms of peanut allergy (van der Veen et al. 1997). Furthermore, among 91 % of those with a discrepancy between specific IgE to peanut and SPT, IgE against CCDs could be detected (van der Veen et al. 1997). In some of these patients, almost complete inhibition of IgE to peanut (as measured by competitive radioallergosorbent test) was possible with CCD. In addition, cross-reactive IgE directed against CCDs in this study was demonstrated to have poor biological activity (van der Veen et al. 1997).

In another study, about 42 % of pollen-allergic European patients were found to have specific IgE to the CCD marker bromelain without skin reactivity to this molecule (Mari et al. 1999). In line with these observations, 23 % of a large group of 1,831 subjects with symptoms of allergic respiratory disease were IgE sensitized to bromelain without SPT to the same molecule (Mari 2002). Taken together, these studies demonstrate the role of anti-CCD IgE in false-positive in vitro allergy test responses. One could argue that this lack of biological activity was due to the fact that bromelain is substituted with a single IgE-binding glycan, making effective cross-linking highly unlikely. The most convincing proof of the complete lack of clinical relevance of CCD-specific IgE was reported by Mari et al. (2008) in a study where grass pollen-allergic patients with high titers of CCD-specific IgE were skin-tested and subjected to an oral challenge with human lactoferrin expressed in rice kernels. This transgenic molecule was substituted with multiple IgE-binding glycans, but both SPT and oral challenge were completely negative (Mari et al. 2008). This study shows that a single IgE-binding glycan is not the reason for lack of activity of IgE to CCD.

With regard to helminth-induced IgE cross-reactivity, a recent study among schoolchildren (aged 5–16 years) in Ghana, West Africa, demonstrated how cross-reactivity between helminth antigens and allergens can affect IgE sensitization patterns and clinical expression of allergy (Amoah et al. 2013b). In this study, the overall prevalence of peanut–IgE sensitization was 17.5 % (233 out of 1,328). However, none of the peanut–IgE-sensitized children had either SPT reactivity to peanut or much reported adverse reactions to peanut. In this study, the presence of *S. haematobium* infection was positively associated with an increased risk of having peanut-specific IgE. In a subset of this study population, both the CCD marker bromelain and *S. haematobium*-soluble egg antigen (SEA) inhibited IgE binding to peanut extract. This study also showed that peanut-specific IgE was strongly correlated with CCD-specific IgE. Furthermore, these results indicate that much of IgE to peanut in Ghanaian children could be directed against CCD which is also present in the schistosome SEA. In addition, basophil histamine release assays demonstrated that the IgE directed against peanut in this population had low biological activity (Amoah et al. 2013b).

This study provides a model which proposes that parasite-induced IgE against CCDs that are carried by parasites might account for high IgE levels to food allergens, and the finding that this IgE does not lead to reactivity to allergenic extracts either in vitro (in basophil release assay) or in vivo (in skin prick testing) further confirms that these IgEs to CCDs are clinically irrelevant (Amoah et al. 2013a).

Although a number of studies have demonstrated that IgE antibodies directed against CCDs are not of clinical relevance, IgE direct against the mammalian carbohydrate epitope galactose-α-1,3-galactose (alpha-gal) has been linked to anaphylactic reactions.

The first cases of anaphylactic reactions associated with alpha-gal were among cancer patients in the southeastern USA receiving therapy with the monoclonal antibody cetuximab (Chung et al. 2008). This antibody carries alpha-gal structure. These patients were shown to have pre-existing IgE against alpha-gal, and immediate-onset anaphylaxis followed the first infusions of cetuximab (Chung et al. 2008). Further analysis indicated that IgE directed against alpha-gal was possibly induced by the lone star tick *Amblyomma americanum* commonly found in the southeastern USA (Commins et al. 2011). Aside from immediate-onset anaphylactic reactions, delayed onset reactions 3–6 h following the consumption of mammalian meat have also been linked to IgE directed against alpha-gal, present on this food (Commins and Platts-Mills 2013).

In addition, positive IgE responses to alpha-gal have been observed in samples from children living in helminth-endemic areas of Ecuador and Kenya (Commins and Platts-Mills 2013). The involvement of helminths in the induction of IgE responses to alpha-gal has been indicated by an investigation conducted in Zimbabwe. In this study, IgE responses to alpha-gal were measured in urban cat-allergic patients as well as in rural helminth-infected subjects (Arkestal et al. 2011). The study observed that 85 % of the parasite-infected group had IgE against alpha-gal and 66 % had IgE against the cat allergen Fel d 5 found in cat dander extract which has been demonstrated to have alpha-gal epitopes (Gronlund et al. 2009). Moreover, in this study, IgE to alpha-gal and IgE to Fel d 5 were highly correlated. Among the urban cat-allergic patients, only a few had IgE responses to Fel d 5 and alpha-gal, while 74 % had responses to the recombinant form of the cat allergen Fel d 1. By contrast, only two of 47 of the parasite-infected had IgE to Fel d 1 which lacks alpha-gal epitopes. Taken together, these observations suggest that in helminth-endemic areas, the IgE to alpha-gal may not be clinically relevant. However, given that no information was collected on reactions to mammalian meat in the helminth-endemic areas (Commins and Platts-Mills 2013), additional in-depth studies are needed to assess the prevalence of sensitization to alpha-gal in different populations worldwide and the relationship between IgE sensitization to this oligosaccharide and clinical outcomes.

3.2 Peptide Cross-Reactivity and Helminths

Cross-reactions between allergens from invertebrates such as mite and snail; cockroach and ascaris; mite, shrimp, and cockroach; and mites and schistosomes (Aalberse et al. 2001) have been reported and involve protein cross-reactivity. Three of the proteins that are involved in these examples are tropomyosin, glutathione S-transferase (GST), and paramyosin (Aalberse et al. 2001).

Tropomyosins are proteins involved in the contraction of muscle cells along with actin and myosin (Arruda and Santos 2005). Not only are tropomyosins major allergens of seafood, mite, and cockroach but are also highly immunogenic helminth proteins (Sereda et al. 2008). Tropomyosins from invertebrates are strong inducers of IgE antibody responses in human (Jenkins et al. 2007). Santiago et al. (2011) demonstrated that there was 72 % identity at the amino acid level between the tropomyosin from the filarial parasite *Onchocerca volvulus* (OvTrop) and the house dust mite tropomyosin Der p 10 (Santiago et al. 2011). A strong correlation between specific IgE to Der p 10 and IgE to OvTrop was shown. In addition, histamine release from basophils sensitized with the sera of individuals IgE positive to Der p 10 could be triggered by either the OvTrop or Der p 10. It is, however, important to realize that such biological activity is not proof of clinical allergy.

In the study by Mari et al. (2008) with transgenic lactoferrin, histamine release was also reported at relatively high protein concentrations, but SPT and oral challenge with the same molecule were negative. The study by Santiago et al. (2011) does, however, confirm that the anti-tropomyosin antibodies induced in filarial infection are cross-reactive with those allergenic tropomyosins of invertebrates (mite) that may affect sensitization and regulation of allergic reactivity. As expected, no clinical mite allergy was reported for the subjects studied, indicating that these cross-reactive responses are of no clinical relevance.

In another investigation, Santos et al. (2008) showed that the predicted structure of *A. lumbricoides* tropomyosin was similar to that of *Periplaneta americana* tropomyosin. The same study compared IgE responses to these proteins in Brazilian children aged 3–6 years living in a helminth-endemic area and cockroach-allergic patients aged 2–52 years also from Brazil (Santos et al. 2008). A strong correlation was also found for IgE antibodies to tropomyosin from *A. lumbricoides* and from *P. Americana* in sera from both populations. Seventy-six percent (90 out of 119) of subjects from the parasite-endemic area had positive IgE antibodies against cockroach tropomyosin without allergy to cockroach (Santos et al. 2008). In line with this study, Acevedo et al. (2009) have also demonstrated high allergenic cross-reactivity between *Blomia tropicalis* tropomyosin (Blo t 10) and *Ascaris* tropomyosin in Colombian asthmatic patients (Acevedo et al. 2009).

The glutathione S-transferases (GSTs) are detoxification enzymes found in most living organisms (Sheehan et al. 2001). The important known sources are cockroaches, house dust mites, and molds; however, GSTs from invertebrates including helminths are known to be strong inducers of IgE (Acevedo et al. 2013). Moreover, *Blattella germanica* GST caused positive immediate skin tests in cockroach-allergic asthmatic patients, suggesting that GST from cockroach is a clinically relevant allergen (Arruda et al. 1997). Regarding helminth-induced IgE cross-reactivity, Santiago et al. (2012) showed that the GSTs from the filarial worm *Wuchereria bancrofti* (WbGST) and cockroach GST (Bla g 5) were 30 % identical at the amino acid with marked similarity in the *N*-terminal region (Santiago et al. 2012). Interestingly, mice infected with *Heligmosomoides bakeri*, a parasite that contains a GST that was 32 % identical to Bla g 5, developed immediate hypersensitivity reaction in the skin to cockroach GST (Bla g 5), suggesting that some parasite-induced cross-reactivity may induce in vivo reactivity to the cross-reactive allergen in a common allergen source like house dust mite (Santiago et al. 2012).

Paramyosin is another allergen family from invertebrate muscle that is targeted in IgE responses against helminths (Fitzsimmons et al. 2014). A study among patients reporting symptoms of allergy and *ascaris*-infected subject in Philippines showed evidence of cross-reactivity between paramyosin from mite (*B. tropicalis*) and paramyosin from *A. lumbricoides* (Valmonte et al. 2012). This study observed that IgE to mite extract among allergic patients can be inhibited, up to 92 %, by *ascaris* antigen, while mite extract could inhibit up to 54 % of *Ascaris*-sIgE among *Ascaris*-infected subjects. Of note, IgE responses to the recombinant form of the paramyosin *Blomia* allergen (Blo t 11) were seen in 80 % of allergic patients and 46 % of *Ascaris*-infected subjects (Valmonte et al. 2012).

In general, IgE cross-reactivity between helminth antigens and allergens demonstrates the limits to diagnostic value of examining IgE responses to whole allergen extracts in helminth-endemic populations. Establishing the molecular basis of cross-reactivity between helminths and common allergen sources is essential to evaluate whether sensitization to the latter is true primary sensitization or cross-reactivity induced by helminths.

4 Component-Resolved Diagnosis (CRD) in Allergy Diagnosis

For the past few decades, in vitro allergy diagnostics has been largely based on the detection of specific IgE to whole extracts comprised of allergenic and non-allergenic components (Treudler and Simon 2013). However, this approach has been problematic since the allergenic content of whole extracts is often difficult to standardize and also the specific allergic reaction inducing components in whole allergen extracts can be hard to identify (Valenta et al. 1999).

Such issues in in vitro allergy diagnostics led to the development of CRD in which purified natural or recombinant allergens are used to detect IgE sensitization to individual allergen molecules (Treudler and Simon 2013).

The use of molecular techniques and recombinant DNA technology has allowed the sequencing, synthesizing, and cloning of allergenic proteins leading to the production of recombinant allergens for CRD (Gadisseur et al. 2011). Recombinant allergens can be generated as defined molecules with consistent quality and without biological variation (Valenta and Niederberger 2007).

The molecular biological techniques underlying CRD were initially employed for the determination of the primary structures and molecular identities of allergens (Valenta et al. 1999). The sequence analysis of allergens allowed the identification of structurally related allergens and also revealed how closely linked cross-reactive molecules may not be differentiated by the immune system (Valenta et al. 1999). CRD involves the use of specific marker allergens to diagnose real sensitization toward a particular allergen source and to discriminate from sensitization to CCD or other homologous allergens (De Knop et al. 2010). It also allows the differentiation between clinically important and irrelevant specific IgEs (Treudler and Simon 2013).

In terms of the application of CRD to research in helminth-endemic populations, the study on peanut allergy among Ghanaian schoolchildren found that in a subset of study subjects with elevated IgE to whole peanut allergen, responses to recombinant forms of the major peanut allergens (rAra h 1, 2 & 3) were generally very low (Amoah et al. 2013b). In addition, among Brazilian children living in a helminth-endemic urban area, Carvalho et al. (2013) evaluated the use of IgE responses to *B. tropicalis* allergens (rBlo t 5 and rBlo t 21) in improving the specificity of determining mite allergy in this population. This study showed that the assays using recombinant allergens exhibited lower IgE cross-reactivity with *A. lumbricoides* antigens and therefore conferred higher specificity in detecting genuine mite IgE sensitization than crude mite extract.

4.1 Microarray

In recent years, microarray biochips have been developed to allow the simultaneous measurement of specific IgE to multiple recombinant and natural allergen components using a small amount of serum. These microarray biochips are increasingly being used in developed countries to provide additional information on IgE profiles of polysensitized allergic patients to improve the management of their conditions.

In a previously published cohort of children in Indonesia, high levels of IgE to house dust mite were found, but this did not translate into SPT reactivity (Hamid et al. 2013). In the study, helminth-induced IgE cross-reactivity was implicated as a possible explanation for the elevated levels of clinically irrelevant allergen-specific IgE. In a subset of these children, the specific IgE to *Dermatophagoides pteronyssinus* (Der p) determined by the ImmunoCAP method (sensitization cutoff ≥0.35 kUA/L) was compared to semiquantitative IgE analysis using a commercially available microarray chip (ImmunoCAP ISAC). It was found that the prevalence of IgE sensitization to whole house dust mite extract (Der p) was 74 %, while sensitization to recombinant and natural house dust mite component allergens as assessed by the microarray biochip was only up to 5 % (Hamid et al. unpublished). This investigation also observed that among these same children, the highest IgE reactivity was to natural allergens of Bermuda and Timothy grass pollen and not much to recombinant grass pollen allergens on the chip indicating that these chips can possibly provide valuable additional information when studying sera with less known IgE specificities. For example, as shown in Fig. 2, the microarray slide contains natural forms of major peanut and grass pollen allergens that contain glycan structures as well as recombinant allergen components that are not glycosylated. The microarray technique used for serum sample from a helminth-infected individual compared to a European allergic patient can differentiate between IgE directed against protein structures that may be biologically active and IgE directed against carbohydrate moieties on glycosylated allergens that might be clinically irrelevant.

5 Future Directions

The studies of helminths and IgE have shed much light but also uncertainty about different aspects of the association between IgE, SPT, and allergy symptoms. Further research into this area should consider the importance of refining and preparing new diagnostic methods for the developing world where allergies are

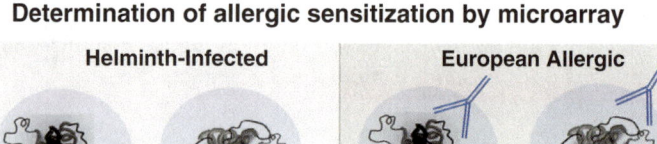

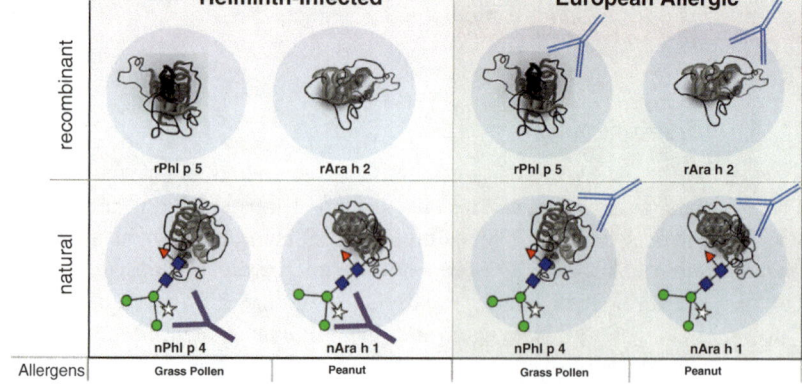

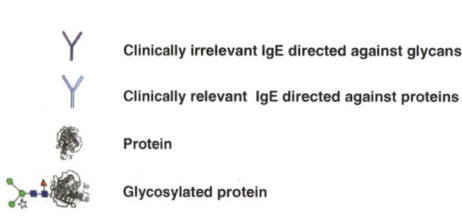

Fig. 2 An illustration of typical component-resolved diagnosis results generated using the ImmunoCAP ISAC™ microarray. The microarray slide shown contains natural grass pollen (nPhl p 4) and recombinant grass pollen (rPhl p 5) as well as natural peanut allergen (nAra h 1) and recombinant peanut allergen (rAra h 2). For the allergic European, the IgE recognizes and binds to the protein structures on both the natural and recombinant allergens for both grass pollen and peanut. This IgE is biologically active and also clinically relevant. In the serum of the helminth-infected subject, elevated levels of IgE are observed that recognize and bind to carbohydrate moieties on the natural allergen components. This IgE does not bind to the protein structures of these components, is clinically irrelevant, and shows poor biological activity

increasing, but the diagnosis is hampered by the complexity of the IgE antibodies. Moreover, further understanding of how IgE cross-reactivity develops and how this affects the biological activity of the antibody is needed. This could help devise interventions to stimulate IgE with poor biological activity which could hamper the development of high-affinity IgE, mast cell degranulation, and allergy symptoms.

References

Aalberse RC, Koshte V, Clemens JG (1981) Immunoglobulin E antibodies that crossreact with vegetable foods, pollen, and Hymenoptera venom. J Allergy Clin Immunol 68:356–364

Aalberse RC, Akkerdaas J, van Ree R (2001) Cross-reactivity of IgE antibodies to allergens. Allergy 56:478–490

Aalberse RC, Stapel SO, Schuurman J et al (2009) Immunoglobulin G4: an odd antibody. Clin Exp Allergy 39:469–477

Acevedo N, Caraballo L (2011) IgE cross-reactivity between Ascaris lumbricoides and mite allergens: possible influences on allergic sensitization and asthma. Parasite Immunol 33:309–321

Acevedo N, Sanchez J, Erler A et al (2009) IgE cross-reactivity between Ascaris and domestic mite allergens: the role of tropomyosin and the nematode polyprotein ABA-1. Allergy 64:1635–1643

Acevedo N, Mohr J, Zakzuk J et al (2013) Proteomic and immunochemical characterization of glutathione transferase as a new allergen of the nematode *Ascaris lumbricoides*. PLoS ONE 8: e78353

Altmann F (2007) The role of protein glycosylation in allergy. Int Arch Allergy Immunol 142:99–115

Amberbir A, Medhin G, Erku W et al (2011) Effects of Helicobacter pylori, geohelminth infection and selected commensal bacteria on the risk of allergic disease and sensitization in 3-year-old Ethiopian children. Clin Exp Allergy 41:1422–1430

Amoah AS, Boakye DA, van Ree R et al (2013a) Parasitic worms and allergies in childhood: insights from population studies 2008–2013. Pediatr Allergy Immunol 25:208–217

Amoah AS, Obeng BB, Larbi IA et al (2013b) Peanut-specific IgE antibodies in asymptomatic Ghanaian children possibly caused by carbohydrate determinant cross-reactivity. J Allergy Clin Immunol 132:639–647

Araujo MI, Lopes AA, Medeiros M et al (2000) Inverse association between skin response to aeroallergens and Schistosoma mansoni infection. Int Arch Allergy Immunol 123:145–148

Arkestal K, Sibanda E, Thors C et al (2011) Impaired allergy diagnostics among parasite-infected patients caused by IgE antibodies to the carbohydrate epitope galactose-alpha 1,3-galactose. J Allergy Clin Immunol 127:1024–1028

Arruda LK, Santos AB (2005) Immunologic responses to common antigens in helminthic infections and allergic disease. Curr Opin Allergy Clin Immunol 5:399–402

Arruda LK, Vailes LD, Platts-Mills TA et al (1997) Induction of IgE antibody responses by glutathione S-transferase from the German cockroach (*Blattella germanica*). J Biol Chem 272:20907–20912

Bager P, Arnved J, Ronborg S et al (2010) Trichuris suis ova therapy for allergic rhinitis: a randomized, double-blind, placebo-controlled clinical trial. J Allergy Clin Immunol 125:123–130

Bethony J, Brooker S, Albonico M et al (2006) Soil-transmitted helminth infections: ascariasis, trichuriasis, and hookworm. Lancet 367:1521–1532

Burke W, Fesinmeyer M, Reed K et al (2003) Family history as a predictor of asthma risk. Am J Prev Med 24:160–169

Carvalho EM, Bastos LS, Araujo MI (2006) Worms and allergy. Parasite Immunol 28:525–534

Carvalho KA, de Melo-Neto OP, Magalhaes FB et al (2013) Blomia tropicalis Blo t 5 and Blo t 21 recombinant allergens might confer higher specificity to serodiagnostic assays than whole mite extract. BMC Immunol 14:11

Chung CH, Mirakhur B, Chan E et al (2008) Cetuximab-induced anaphylaxis and IgE specific for galactose-alpha-1,3-galactose. N Engl J Med 358:1109–1117

Commins SP, Platts-Mills TA (2013) Delayed anaphylaxis to red meat in patients with IgE specific for galactose alpha-1,3-galactose (alpha-gal). Curr Allergy Asthma Rep 13:72–77

Commins SP, James HR, Kelly LA et al (2011) The relevance of tick bites to the production of IgE antibodies to the mammalian oligosaccharide galactose-alpha-1,3-galactose. J Allergy Clin Immunol 127:1286–1293

Cooper PJ (2009) Interactions between helminth parasites and allergy. Curr Opin Allergy Clin Immunol 9:29–37

Cooper PJ, Chico ME, Bland M et al (2003a) Allergic symptoms, atopy, and geohelminth infections in a rural area of Ecuador. Am J Respir Crit Care Med 168:313–317

Cooper PJ, Chico ME, Rodrigues LC et al (2003b) Reduced risk of atopy among school-age children infected with geohelminth parasites in a rural area of the tropics. J Allergy Clin Immunol 111:995–1000

Cooper PJ, Chico ME, Rodrigues LC et al (2004) Risk factors for atopy among school children in a rural area of Latin America. Clin Exp Allergy 34:845–852

Cooper PJ, Chico ME, Vaca MG et al (2006) Effect of albendazole treatments on the prevalence of atopy in children living in communities endemic for geohelminth parasites: a cluster-randomised trial. Lancet 367:1598–1603

Dagoye D, Bekele Z, Woldemichael K et al (2003) Wheezing, allergy, and parasite infection in children in urban and rural Ethiopia. Am J Respir Crit Care Med 167:1369–1373

De Knop KJ, Bridts CH, Verweij MM et al (2010) Component-resolved allergy diagnosis by microarray: potential, pitfalls, and prospects. Adv Clin Chem 50:87–101

Feary JR, Venn AJ, Mortimer K et al (2010) Experimental hookworm infection: a randomized placebo-controlled trial in asthma. Clin Exp Allergy 40:299–306

Feary J, Britton J, Leonardi-Bee J (2011) Atopy and current intestinal parasite infection: a systematic review and meta-analysis. Allergy 66:569–578

Fitzsimmons CM, Falcone FH, Dunne DW (2014) Helminth allergens, parasite-specific IgE, and its protective role in human immunity. Front Immunol 5:61

Flohr C, Tuyen LN, Lewis S et al (2006) Poor sanitation and helminth infection protect against skin sensitization in Vietnamese children: a cross-sectional study. J Allergy Clin Immunol 118:1305–1311

Flohr C, Tuyen LN, Quinnell RJ et al (2010) Reduced helminth burden increases allergen skin sensitization but not clinical allergy: a randomized, double-blind, placebo-controlled trial in Vietnam. Clin Exp Allergy 40:131–142

Gadisseur R, Chapelle JP, Cavalier E (2011) A new tool in the field of in-vitro diagnosis of allergy: preliminary results in the comparison of ImmunoCAP(c) 250 with the ImmunoCAP(c) ISAC. Clin Chem Lab Med 49:277–280

Godfrey RC (1975) Asthma and IgE levels in rural and urban communities of the Gambia. Clin Allergy 5:201–207

Gronlund H, Adedoyin J, Commins SP et al (2009) The carbohydrate galactose-alpha-1,3-galactose is a major IgE-binding epitope on cat IgA. J Allergy Clin Immunol 123(5):1189–1191

Hamid F, Wiria AE, Wammes LJ et al (2013) Risk factors associated with the development of atopic sensitization in Indonesia. PLoS ONE 8:e67064

Jenkins JA, Breiteneder H, Mills EN (2007) Evolutionary distance from human homologs reflects allergenicity of animal food proteins. J Allergy ClinImmunol 120:1399–1405

Jouvin MH, Kinet JP (2012) *Trichuris suis* ova: testing a helminth-based therapy as an extension of the hygiene hypothesis. J Allergy Clin Immunol 130:3–10

Lancet editorial (1976) Editorial: IgE, parasites, and allergy. Lancet 24:894–895

Larson D, Cooper PJ, Hubner MP et al (2012a) Helminth infection is associated with decreased basophil responsiveness in human beings. J Allergy Clin Immunol 130:270–272

Larson D, Hubner MP, Torrero MN et al (2012b) Chronic helminth infection reduces basophil responsiveness in an IL-10-dependent manner. J Immunol 188:4188–4199

Lynch NR, Hagel I, Perez M et al (1993) Effect of anthelmintic treatment on the allergic reactivity of children in a tropical slum. J Allergy Clin Immunol 92:404–411

Lynch NR, Palenque M, Hagel I et al (1997) Clinical improvement of asthma after anthelminthic treatment in a tropical situation. Am J Respir Crit Care Med 156:50–54

Lynch NR, Hagel IA, Palenque ME et al (1998) Relationship between helminthic infection and IgE response in atopic and nonatopic children in a tropical environment. J Allergy Clin Immunol 101:217–221

Macaubas C, Sly PD, Burton P et al (1999) Regulation of T-helper cell responses to inhalant allergen during early childhood. Clin Exp Allergy 29:1223–1231

MacGlashan DW Jr, Bochner BS, Adelman DC et al (1997) Down-regulation of Fc(epsilon)RI expression on human basophils during in vivo treatment of atopic patients with anti-IgE antibody. J Immunol 158:1438–1445

Mari A (2002) IgE to cross-reactive carbohydrate determinants: analysis of the distribution and appraisal of the in vivo and in vitro reactivity. Int Arch Allergy Immunol 129:286–295

Mari A, Iacovacci P, Afferni C et al (1999) Specific IgE to cross-reactive carbohydrate determinants strongly affect the in vitro diagnosis of allergic diseases. J Allergy Clin Immunol 103:1005–1011

Mari A, Ooievaar-de HP, Scala E et al (2008) Evaluation by double-blind placebo-controlled oral challenge of the clinical relevance of IgE antibodies against plant glycans. Allergy 63:891–896

Mitre E, Norwood S, Nutman TB (2005) Saturation of immunoglobulin E (IgE) binding sites by polyclonal IgE does not explain the protective effect of helminth infections against atopy. Infect Immun 73:4106–4111

Mpairwe H, Webb EL, Muhangi L et al (2011) Anthelminthic treatment during pregnancy is associated with increased risk of infantile eczema: randomised-controlled trial results. Pediatr Allergy Immunol 22:305–312

Obeng BB, Amoah AS, Larbi IA et al (2014) Schistosoma infection is negatively associated with mite atopy, but not wheeze and asthma in Ghanaian Schoolchildren. Clin Exp Allergy 44:965–975

Obihara CC, Beyers N, Gie RP et al (2006) Respiratory atopic disease, Ascaris-immunoglobulin E and tuberculin testing in urban South African children. Clin Exp Allergy 36:640–648

Palmer LJ, Celedon JC, Weiss ST et al (2002) Ascaris lumbricoides infection is associated with increased risk of childhood asthma and atopy in rural China. Am J Respir Crit Care Med 165:1489–1493

Pawankar R, Canonica GW, Holgate ST et al (2011) Introduction and executive summary: allergic diseases as a global public health issue. In: Pawankar R et al (eds) World Allergy Organization (WAO) white book on allergy. pp 11–20

Perzanowski MS, Ng'ang'a LW, Carter MC et al (2002) Atopy, asthma, and antibodies to Ascaris among rural and urban children in Kenya. J Pediatr 140:582–588

Ponte EV, Lima F, Araujo MI et al (2006) Skin test reactivity and Der p-induced interleukin 10 production in patients with asthma or rhinitis infected with Ascaris. Ann Allergy Asthma Immunol 96:713–718

Priftanji A, Strachan D, Burr M et al (2001) Asthma and allergy in Albania and the UK. Lancet 358:1426–1427

Royer B, Varadaradjalou S, Saas P et al (2001) Inhibition of IgE-induced activation of human mast cells by IL-10. Clin Exp Allergy 31:694–704

Santiago HC, Bennuru S, Boyd A et al (2011) Structural and immunologic cross-reactivity among filarial and mite tropomyosin: implications for the hygiene hypothesis. J Allergy Clin Immunol 127:479–486

Santiago HC, LeeVan E, Bennuru S et al (2012) Molecular mimicry between cockroach and helminth glutathione S-transferases promotes cross-reactivity and cross-sensitization. J Allergy Clin Immunol 130:248–256

Santos AB, Rocha GM, Oliver C et al (2008) Cross-reactive IgE antibody responses to tropomyosins from *Ascaris lumbricoides* and cockroach. J Allergy Clin Immunol 121:1040–1046

Satoguina JS, Adjobimey T, Arndts K et al (2008) Tr1 and naturally occurring regulatory T cells induce IgG4 in B cells through GITR/GITR-L interaction, IL-10 and TGF-beta. Eur J Immunol 38:3101–3113

Scrivener S, Yemaneberhan H, Zebenigus M et al (2001) Independent effects of intestinal parasite infection and domestic allergen exposure on risk of wheeze in Ethiopia: a nested case-control study. Lancet 358:1493–1499

Sereda MJ, Hartmann S, Lucius R (2008) Helminths and allergy: the example of tropomyosin. Trends Parasitol 24:272–278

Sheehan D, Meade G, Foley VM et al (2001) Structure, function and evolution of glutathione transferases: implications for classification of non-mammalian members of an ancient enzyme superfamily. Biochem J 360:1–16

Smits HH, Hammad H, van Nimwegen M et al (2007) Protective effect of *Schistosoma mansoni* infection on allergic airway inflammation depends on the intensity and chronicity of infection. J Allergy Clin Immunol 120:932–940

Smits HH, Everts B, Hartgers FC et al (2010) Chronic helminth infections protect against allergic diseases by active regulatory processes. Curr Allergy Asthma Rep 10:3–12

Supali T, Djuardi Y, Wibowo H et al (2010) Relationship between different species of helminths and atopy: a study in a population living in helminth-endemic area in Sulawesi, Indonesia. Int Arch Allergy Immunol 153:388–394

Treudler R, Simon JC (2013) Overview of component resolved diagnostics. Curr Allergy Asthma Rep 13:110–117

Valenta R, Niederberger V (2007) Recombinant allergens for immunotherapy. J Allergy Clin Immunol 119:826–830

Valenta R, Lidholm J, Niederberger V et al (1999) The recombinant allergen-based concept of component-resolved diagnostics and immunotherapy (CRD and CRIT). Clin Exp Allergy 29:896–904

Valmonte GR, Cauyan GA, Ramos JD (2012) IgE cross-reactivity between house dust mite allergens and Ascaris lumbricoides antigens. Asia Pac Allergy 2:35–44

van den Biggelaar AH, van Ree R, Rodrigues LC et al (2000) Decreased atopy in children infected with *Schistosoma haematobium*: a role for parasite-induced interleukin-10. Lancet 356:1723–1727

van den Biggelaar AH, Lopuhaa C, van Ree R et al (2001) The prevalence of parasite infestation and house dust mite sensitization in Gabonese schoolchildren. Int Arch Allergy Immunol 126:231–238

van der Veen MJ, van Ree R, Aalberse RC et al (1997) Poor biologic activity of cross-reactive IgE directed to carbohydrate determinants of glycoproteins. J Allergy Clin Immunol 100:327–334

van Ree R, Cabanes-Macheteau M, Akkerdaas J et al (2000) Beta(1,2)-xylose and alpha(1,3)-fucose residues have a strong contribution in IgE binding to plant glycoallergens. J Biol Chem 275:11451–11458

von Mutius E (2002) Environmental factors influencing the development and progression of pediatric asthma. J Allergy ClinImmunol 109:S525–S532

Wilson MS, Taylor MD, Balic A et al (2005) Suppression of allergic airway inflammation by helminth-induced regulatory T cells. J Exp Med 202:1199–1212

Wiria AE, Hamid F, Wammes LJ et al (2013) The effect of three-monthly albendazole treatment on malarial parasitemia and allergy: a household-based cluster-randomized, double-blind, placebo-controlled trial. PLoS ONE 8:e57899

Wright V, Bickle Q (2005) Immune responses following experimental human hookworm infection. Clin Exp Immunol 142:398–403

Yazdanbakhsh M, van den Biggelaar A, Maizels RM (2001) Th2 responses without atopy: immunoregulation in chronic helminth infections and reduced allergic disease. Trends Immunol 22:372–377

Yazdanbakhsh M, Kremsner PG, van Ree R (2002) Allergy, parasites, and the hygiene hypothesis. Science 296:490–494

IgE Immunotherapy Against Cancer

Lai Sum Leoh, Tracy R. Daniels-Wells and Manuel L. Penichet

Abstract The success of antibody therapy in cancer is consistent with the ability of these molecules to activate immune responses against tumors. Experience in clinical applications, antibody design, and advancement in technology have enabled antibodies to be engineered with enhanced efficacy against cancer cells. This allows re-evaluation of current antibody approaches dominated by antibodies of the IgG class with a new light. Antibodies of the IgE class play a central role in allergic reactions and have many properties that may be advantageous for cancer therapy. IgE-based active and passive immunotherapeutic approaches have been shown to be effective in both in vitro and in vivo models of cancer, suggesting the potential use of these approaches in humans. Further studies on the anticancer efficacy and safety profile of these IgE-based approaches are warranted in preparation for translation toward clinical application.

L.S. Leoh · T.R. Daniels-Wells (✉) · M.L. Penichet (✉)
Division of Surgical Oncology, Department of Surgery, David Geffen School of Medicine, University of California, Los Angeles, 10833 Le Conte Avenue, CHS 54-140, Box 951782, Los Angeles, CA 90095-1782, USA
e-mail: tdaniels@mednet.ucla.edu

M.L. Penichet
e-mail: penichet@mednet.ucla.edu

M.L. Penichet
Department of Microbiology, Immunology, and Molecular Genetics, University of California, 609 Charles E. Young Dr. East, 1602 Molecular Science Building, Los Angeles, CA 90095, USA

M.L. Penichet
The Jonsson Comprehensive Cancer Center, University of California, 10833 Le Conte Ave, 8-684 Factor Building, Box 951781, Los Angeles, CA 90095, USA

M.L. Penichet
The Molecular Biology Institute, University of California, 611 Charles E. Young Dr., Los Angeles, CA 90095, USA

© Springer International Publishing Switzerland 2015
J.J. Lafaille and M.A. Curotto de Lafaille (eds.), *IgE Antibodies: Generation and Function*, Current Topics in Microbiology and Immunology 388,
DOI 10.1007/978-3-319-13725-4_6

Contents

1 Immunoglobulins and Their Relevance in Cancer	110
1.1 Immunoglobulins	110
1.2 The Structure of Immunoglobulins	111
1.3 Antibodies for Cancer Immunotherapy	111
1.4 The IgE Antibody and Its Receptors	112
1.5 IgE and Cancer	115
2 IgE-based Immunotherapy	120
2.1 Passive Immunotherapy	121
2.2 Active Immunotherapy (Vaccination Approaches)	133
3 Safety Concerns	136
4 Concluding Thoughts	138
References	139

1 Immunoglobulins and Their Relevance in Cancer

1.1 Immunoglobulins

Immunoglobulins, also known as antibodies, were the first characterized molecules involved in specific immune recognition. Antitoxins against tetanus and diphtheria toxins were discovered in the 1890s by Shibasaburo Kitasato and Emil von Behring (Behring and Kitasato 1890). Collaboration between Behring and Paul Ehrlich enabled the production of a standardized, efficient serum therapy for the treatment of diphtheria (Winau and Winau 2002). Subsequently, Paul Ehrlich formulated the concepts of active and passive immunization (Ehrlich 1891) and developed the side chain theory (Ehrlich 1901b), describing receptors that bind distinct toxins on the cell surface with lock-and-key specificity (Winau et al. 2004). These receptors, representing antitoxins or antibodies, are released into the blood (Ehrlich 1901a). Since then, antibodies, with their unique specificity to recognize distinct target molecules (known as antigens), have been utilized to attack tumor cells expressing certain antigens (Sliwkowski and Mellman 2013; Weiner et al. 2010). Introduction of the hybridoma technology (Kohler and Milstein 1975) enabled mass production of mouse monoclonal antibodies with a single specificity. This technology, along with advances in bioengineering, has facilitated the development of chimeric, humanized, and fully human monoclonal antibodies with decreased immunogenicity and enhanced anticancer efficacy that can be used as effective anticancer therapeutics in humans.

1.2 The Structure of Immunoglobulins

Antibodies are composed of 2 identical heavy (H) and 2 identical light (L) chains, exhibiting a H_2L_2 heterotetramer configuration (Janeway et al. 2005a). Each chain has both constant and variable regions. Heavy chains can pair with either kappa (κ) or lambda (λ) light chains. There are 5 different classes of antibodies in humans distinguished by their heavy chain structure denoted by the Greek letters: α (IgA), δ (IgD), ε (IgE), γ (IgG), and μ (IgM). IgD, IgG, and IgE are monomeric antibodies. There are 4 subclasses of IgG (IgG1, IgG2, IgG3, and IgG4), while IgA has 2 subclasses (IgA1 and IgA2). IgG is the main antibody class found in blood and extracellular fluid and protects the body from infection (Janeway et al. 2005a). IgE is associated with type I hypersensitivity (anaphylactic/allergic) reactions. IgM is the first responder to an antigenic challenge, such as an infection, and exists as a pentamer or hexamer. IgA is secreted through body fluids, while IgD (or an IgM monomer) forms the B-cell receptor on the surface of the B cell. The approximate molecular weights of the different classes are as follows: 184 kDa for IgD; 188 kDa for IgE; 146 kDa for IgG1, IgG2, and IgG4; 165 kDa for IgG3 due to an extended hinge region; 160 kDa for both subclasses of monomeric IgA in serum; 390 kDa for secretory dimeric IgA; 970 kDa for pentameric IgM; and 1,140 kDa for hexameric IgM (Janeway et al. 2005a; Murphy 2012).

1.3 Antibodies for Cancer Immunotherapy

As of 2013, 15 antibodies have been approved by the United States Food and Drug Administration (FDA) for the treatment of cancer, with many more undergoing evaluation in clinical trials (Lewin and Thomas 2013; Sliwkowski and Mellman 2013; Cameron and McCormack 2014). Five antibodies target the B-cell marker CD20, including rituximab (Rituxan®, mouse/human chimeric IgG1), the first monoclonal antibody approved for the treatment of cancer (indolent lymphoma) in 1997 (Leget and Czuczman 1998), and a new antibody, obinutuzumab (Gazyva™, humanized IgG1) glycoengineered for higher binding affinity to the FcγRIIIa (Cameron and McCormack 2014). Trastuzumab (Herceptin®), a humanized IgG1 specific for HER2/*neu*, a member of the epidermal growth factor receptor (EGFR) family, was approved for the treatment of HER2/*neu*-positive metastatic breast cancer as a single agent in 1998 (Vogel et al. 2001). Since then, trastuzumab in combination with conventional chemotherapy has also been approved for the treatment of HER2/*neu*-positive breast cancer (Slamon et al. 2001). Cetuximab (Erbitux®), a chimeric mouse/human IgG1 targeting EGFR, combined with chemotherapy was approved for the treatment of wild-type KRAS (Kirsten rat sarcoma viral oncogene homolog) and EGFR-expressing metastatic colorectal cancer (Sliwkowski and Mellman 2013). Multiple mechanisms of action have been described for antibodies targeting tumors including the direct inhibition of target function, complement-mediated cytotoxicity (CDC), antibody-dependent cell-mediated cytotoxicity (ADCC), and antibody-dependent cell-mediated phagocytosis (ADCP). However, Fc-mediated functions, in particular ADCC, are believed to be a major component of their antitumor activity (Campoli et al. 2010; Kellner et al. 2014; Sliwkowski and Mellman 2013; Weiner et al. 2012).

The above-mentioned FDA-approved antibodies targeting cancer cells and other antibodies in development are of the IgG class. IgA molecules have also been evaluated as anticancer agents. An anti-EGFR IgA2 containing the variable regions of cetuximab significantly decreased the number of metastasis in an established immunocompetent lung metastasis model using mouse melanoma cells (B16F10) expressing human EGFR in C57BL/6 human FcαRI transgenic mice (Boross et al. 2013). The effect of this IgA2 lasted a week longer (up to 30 days) than its IgG counterpart cetuximab. IgE is another antibody class that is currently being explored as a potential cancer therapeutic. Studies using IgE in the context of cancer belong to the rapidly growing field of AllergoOncology, which aims to reveal the function of IgE-mediated immune responses against cancer cells in order to enhance the understanding of its biology and to develop novel IgE-based treatment options against malignant diseases (Penichet and Jensen-Jarolim 2010).

1.4 The IgE Antibody and Its Receptors

IgE antibodies have a different structure compared to IgG1 antibodies (Fig. 1). The heavy chain of IgG has 3 constant domains and 1 variable region, while the IgE heavy chain contains an extra constant domain, having a total of 4 constant domains and 1 variable domain. They have similar light chains with 1 variable and 1 constant domain each. The variable regions bind specific antigens, while the constant regions are responsible for effector functions. IgE has 6 N-linked glycosylation sites compared to 1 for IgG. Crystal structure and solution scattering data have revealed a bent structure of the IgE Fc region, which undergoes substantial conformational changes upon binding to Fc epsilon receptor I (FcεRI) (Beavil et al. 1995; Wan et al. 2002). There are 2 receptors for IgE: the high-affinity receptor, FcεRI, and the low-affinity receptor, FcεRII, also known as CD23.

FcεRI expression is abundant on human mast cells (MC) and basophils and is expressed at lower levels on dendritic cells (DC), Langerhans cells (LC), monocytes/macrophages, eosinophils, and platelets (Kinet 1999). It is expressed as an $\alpha\beta\gamma_2$ heterotetramer in MC and basophils and as an $\alpha\gamma_2$ heterotrimer in monocytes and eosinophils (Ravetch and Kinet 1991). The extracellular α subunit is sufficient for IgE binding (Hakimi et al. 1990), while the β and γ subunits are involved in cell signaling (Gould et al. 2003). In the mouse, FcεRI is only expressed in MC and basophils (Kinet 1999) and not in eosinophils (de Andres et al. 1997) and macrophages. However, FcεRI is present on rat eosinophils, monocytes/macrophages, and platelets (Capron and Dessaint 1985; Dombrowicz et al. 2000; Gould and Sutton 2008).

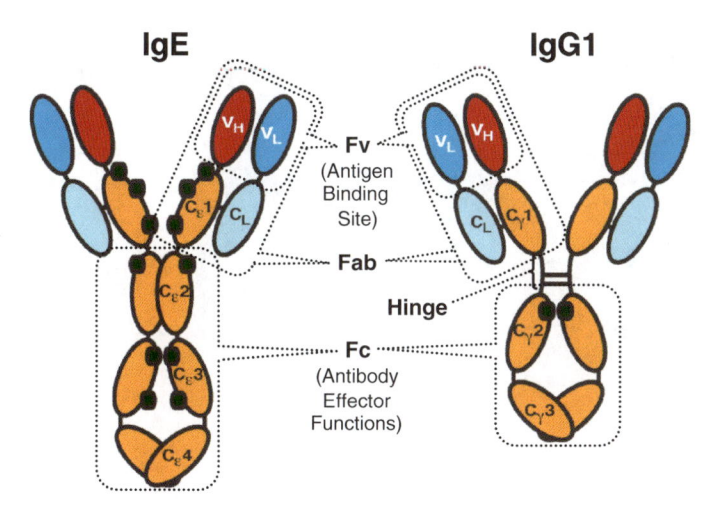

Fig. 1 Schematic diagram comparing the structures of human IgE and IgG1. Antibodies are composed of 2 pairs of identical heavy (H) and light (L) chain proteins linked by disulfide bonds forming H$_2$L$_2$ heterotetramers. The Fab region consists of a constant and a variable domain from each heavy and light chain. The variable region (Fv), composed of both the heavy and light chain variable domains, is responsible for antigen binding and is located at the amino-terminus of the antibody. The remaining constant regions include the Fc portion of the antibody and are responsible for the effector functions. *Black circles* denote *N*-linked glycosylation sites. The hinge region, which provides flexibility, joins the Cγ1 and Cγ2 domains in IgG1. Cε2 replaces the hinge region in IgE. Reprinted from Fig. 7.2 of Daniels et al. (2010) with kind permission from Springer Science and Business Media

IgE also binds to FcεRII (CD23), a type II integral membrane protein with a calcium-dependent lectin domain 'head.' CD23 exists in the membrane as an equilibrium mixture of a monomer and a trimer that is held together by a coiled-coil stalk (Gould et al. 2003). There are 2 isoforms of human CD23. The CD23a isoform is expressed on antigen-activated B cells prior to differentiation into antibody-secreting plasma cells (Gould et al. 2003). CD23a is involved in IgE antibody-dependent antigen endocytosis, processing, and presentation. CD23b is expressed on monocytes and eosinophils following interleukin-4 (IL-4) stimulation (Yokota et al. 1992). Murine and human CD23 show similar structures; however, their pattern of expression is different. Human CD23 is expressed on monocytes/macrophages, lymphocytes, eosinophils, follicular DC, LC, and platelets (Delespesse et al. 1991), while murine CD23 is present on B cells and a subset of CD8 T cells (Mathur et al. 1988).

IgE antibodies play a central role in triggering the allergic immune response (type I hypersensitivity) when exposed to allergens, such as pollen, which leads to MC and/or basophil degranulation and acute inflammation at the site of allergen challenge (Fig. 2a) (Gould and Sutton 2008; Gould et al. 2003). Allergen sensitization results in the production of allergen-specific IgE, which then coat the surface of FcεRI-bearing cells. Re-exposure to the same allergen triggers MC and basophil

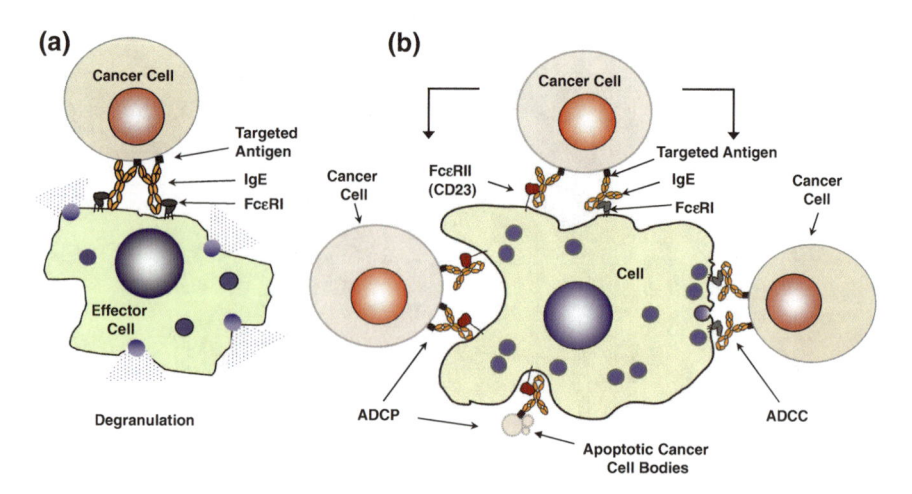

Fig. 2 Scheme of possible IgE-mediated interactions between effector cells and targeted tumor cells. a IgE can trigger degranulation upon cross-linking of FcεRI on the surface of effector cells, such as MC, in the presence of tumor cells expressing high density of the targeted antigen. This degranulation is the type I hypersensitivity (allergic) reaction also triggered in the presence of allergens, which in the case of targeted cancer antigens may result in cytotoxicity and cancer cell death. b The Fc region of the IgE specific for a tumor antigen interacts with the FcεRI or FcεRII/CD23 expressed on the surface of effector cells such as monocytes/macrophages and may trigger cell killing through ADCC or ADCP. Antibodies bound to apoptotic cells or apoptotic bodies are also expected to trigger phagocytosis. Adapted from Fig. 7.3 of Daniels et al. (2010) with kind permission from Springer Science and Business Media

degranulation due to receptor cross-linking, resulting in recruitment of inflammatory cells. The early phase reaction of the type I hypersensitivity response occurs within minutes of exposure to an allergen, releasing preformed mediators including histamine, heparin, lipid mediators, proteases, chemokines, and cytokines from cytoplasmic granules (Galli and Tsai 2010; Karagiannis et al. 2012). Since engagement of multiple FcεRIs is necessary for cross-linking (Kanner and Metzger 1983; Scholl et al. 2005a; Segal et al. 1977), degranulation can only be triggered by multi-epitopic antigens. The late phase reaction occurs after early phase symptoms have diminished and can last up to several weeks. Recruited inflammatory cells include neutrophils, eosinophils, basophils, monocytes/macrophages, and T cells (Ying et al. 1999). Together, these factors lead to tissue damage and an acute inflammatory response. Additionally, IgE-dependent antigen presentation by antigen-presenting cells (APC) such as DC, LC, and macrophages may lead to a secondary immune response. Upregulation of FcεRI by IgE (Karagiannis et al. 2007; MacGlashan et al. 1999) and CD23 by Interleukin-4 (IL-4) (Karagiannis et al. 2007; Spiegelberg 1990) on effector cells may also enhance these pathways. IgE has also been thought to play a role in parasitic infections (Capron et al. 1999; Capron and Capron 1994; Dunne et al. 1992; Gould and Sutton 2008; Obata-Ninomiya et al. 2013); however, this role has been questioned (Cooper et al. 2008; Watanabe et al. 2005).

1.5 IgE and Cancer

1.5.1 Clinical and Epidemiological Observations

There is evidence suggesting that IgE may play a role in antitumor immunity. An immunohistochemistry study on the distribution of the immunoglobulin classes in head and neck cancer revealed the IgE antibody to be the most abundant class in the tumor tissues (Neuchrist et al. 1994). Significantly elevated levels of tumor-specific serum IgE have been observed in patients with pancreatic cancer compared to healthy controls (Fu et al. 2008). Importantly, these IgE antibodies induced ADCC against pancreatic cancer cells (Fu et al. 2008). In multiple myeloma patients, higher levels of polyclonal IgE in non-allergic individuals have been associated with lower disease incidence and longer survival (Matta et al. 2007). Additionally, the presence of IgE effector eosinophils in the blood or peritumoral infiltrates of both hematological and solid malignancies has been linked to favorable prognosis, particularly in solid tumors (Gatault et al. 2012; Munitz and Levi-Schaffer 2004). The frequent state of eosinophil degranulation in close proximity to the tumor (Caruso et al. 2011) suggests their cytotoxic potential in eliminating cancer cells. Furthermore, epidemiological studies have found inverse correlations between the incidence of allergies and the risk of certain malignancies such as glioma, non-Hodgkin lymphoma (NHL), and childhood leukemia, especially acute lymphoblastic leukemia (Josephs et al. 2013; Martínez-Maza et al. 2010; Turner et al. 2006). Taken together, the above studies suggest a potential natural role of IgE in cancer immune surveillance, at least in certain malignancies.

1.5.2 Advantages of IgE Compared to IgG

Affinity for FcɛRs

IgE antibodies have several properties that make them attractive as cancer therapeutics. One of these properties is the high affinity of IgE for its FcɛRs. The binding affinity of human IgE for FcɛRI ($K_a = 10^{10}$ M^{-1}) is two orders of magnitude higher than that of IgG for its high affinity receptor FcγRI ($K_a = 10^8$ M^{-1}) (Daniels et al. 2010; Gould et al. 2003; Janeway et al. 2005d). This high affinity reflects the slow dissociation rate with a half-life of about 20 h for IgE on the receptor (McDonnell et al. 2001), with residence life on MC in tissues extended up to months (Achatz et al. 2010) as a result of restricted diffusion and rebinding to cell receptors (Gould et al. 2003). The affinity of human IgE for the CD23 monomer is around 10^6–10^7 M^{-1}, while the formation of the CD23 trimer leads to a 10-fold higher affinity for human IgE with a K_a of about 10^8–10^9 M^{-1}, which is equivalent to the affinity of IgG1 for FcγRI (Conrad 1990; Gould and Sutton 2008; Hibbert et al. 2005; Kilmon et al. 2001; McCloskey et al. 2007; Ravetch and Kinet 1991).

Human Effector Cells

The effector cells involved in the allergic reaction are unique and thus, are a potential advantage of IgE-based cancer therapies. Several mechanisms may be involved in the potential anticancer activity of tumor-targeted IgE (Fig. 2). IgE on the surface of these effector cells would result in cells better suited to target cancer cells. The use of tumor-specific IgE could arm effector cells to recognize tumor antigens, enhancing local tumor cell killing and inducing a secondary anti-tumor response, possibly inducing long-term antitumor immunity.

MC are tissue-based immune cells of hematopoietic origin located primarily in association with blood vessels and at epithelial surfaces (Stone et al. 2010). MC constitutively express FcεRI on their surface and are central effector cells in IgE-mediated allergic reactions. IgE-mediated antigen activation of MC triggers the type I hypersensitivity reaction releasing inflammatory mediators such as histamine, heparin, proteases, chemokines, cytokines, and lipid mediators (Galli and Tsai 2010). MC can potentially mediate pro-inflammatory or immunosuppressive functions (Galli et al. 2008). It has been observed that the microanatomical location of MC within the tumor microenvironment is important and can lead to different outcomes. A high density of MC is associated with unfavorable prognosis in malignancies such as pancreatic and colorectal cancer, Hodgkin lymphoma (HL), and aggressive skin cancer such as malignant melanoma (Dalton and Noelle 2012). However, in other human cancers such as breast and non-small-cell lung cancer, high MC density in tumors is associated with favorable prognosis (Dalton and Noelle 2012). Peritumoral MC in prostate cancer contributes to angiogenesis and tumor invasion into surrounding tissues and is significantly associated with unfavorable prognosis and poor survival (Johansson et al. 2010). However, intratumoral MC are associated with favorable prognosis in prostate cancer (Dalton and Noelle 2012). In fact, the cytotoxic effect of MC in breast cancer was demonstrated by histological images of degranulating MC surrounding dying tumor cells (della Rovere et al. 2007). The release of cytotoxic compounds from MC including tumor necrosis factor-α (TNF-α) leads to inflammation, cytotoxic T lymphocyte (CTL) responses, and tumor destruction (Wasiuk et al. 2010). Importantly, MC degranulation may also cause inhibition of the function of regulatory T cells (Tregs), which may enhance the immune response (de Vries et al. 2009). Therefore, administration of tumor-specific IgE is expected to enhance the protective effect of MC in tumor types that show protection, as well as to 're-educate' MC to target tumor antigens in the other tumor types where MC are not protective, triggering an 'explosive' degranulation leading to tumor destruction.

Basophils develop from CD34$^+$ progenitors in the bone marrow, where they differentiate and mature before relocating to the circulation (Stone et al. 2010). They constitute less then 1 % of granulocytes in healthy humans and are involved in allergic/anaphylactic (type I hypersensitivity) reactions. Similar to MC, they express FcεRI, secrete T_H2 (T helper 2) cytokines, and release histamine, lipid mediators, chemokines, and cytokines after activation (Stone et al. 2010). Taken together, the above properties suggest that the use of tumor-specific IgE could potentially be manipulated to direct basophil degranulation at localized tumor sites resulting in anti-tumor activity.

Eosinophils are granulocytes that originate in bone marrow and are named for their granules colored by the acidic stain eosin (Janeway et al. 2005c). Most eosinophils reside in mucosal tissues and only very small numbers of these cells are normally present in the circulation, implying a likely role in the defense against invading organisms. Blood and tissue eosinophilia are hallmarks of helminth infection, allergy, asthma, eosinophilic gastrointestinal disorders, and other rare disorders (Stone et al. 2010). Upon IgE activation, eosinophils can release cytotoxic mediators such as eosinophil cationic protein (ECP), major basic protein (MBP), eosinophil peroxidase (EPO), eosinophil-derived neurotoxin (EDN), and free radicals. Activation also induces the synthesis of a variety of chemical mediators such as prostaglandins, leukotrienes, cytokines, and chemokines, which amplify the inflammatory response by activating epithelial cells and recruiting and activating more eosinophils and other leukocytes (Janeway et al. 2005c). Eosinophils have been observed to infiltrate malignant tissues, a phenomenon known as tumor-associated tissue eosinophilia (TATE), which has been associated with significantly better prognosis in esophageal squamous cell carcinoma as well as colorectal, oral, and gastric cancer, (Cuschieri et al. 2002; Dorta et al. 2002; Fernandez-Acenero et al. 2000; Ishibashi et al. 2006). However, it has been associated with poor prognosis in HL (von Wasielewski et al. 2000). Recent reports suggest that eosinophils are involved in tumoricidal activity. In fact, TNF-α, together with granzyme A, was shown to direct tumoricidal activity of eosinophils against colon carcinoma cells (Legrand et al. 2010). Clearance of melanoma metastases by CD4 T_H2 cells was associated with a large influx of eosinophils into the tumor, as detected by immunohistochemistry staining of degranulating eosinophils and its cytotoxic protein MBP in lung metastatic sections (Mattes et al. 2003). Eosinophil infiltration of tumors has been reported to be an early and persistent inflammatory host response, as demonstrated in C57/BL6 J mice challenged subcutaneously (s.c.) with B16-F10 mouse melanoma cells (Cormier et al. 2006). In addition, the use of IL-5 transgenic mice, containing elevated levels of circulating eosinophils, demonstrated significant reduction in tumor establishment and growth, which directly correlated with the recruitment of eosinophils and other myeloid cells to the tumor and surrounding connective tissue (Simson et al. 2007). The above observations indicate that eosinophils may have a protective role in the context of tumor eradication. Thus, human eosinophils, which express both FcεRs, should be able to interact with tumor-targeted IgE to enhance these anti-tumor effects.

Monocytes develop from dividing monoblasts in the bone marrow and are released into the bloodstream as nondividing cells (Lee et al. 2013). Differentiation of monocytes in tissues gives rise to macrophages, phagocytes that are present essentially in all tissues and function in both innate and adaptive immunity (Cook and Hagemann 2013; Ovchinnikov 2008). Classical M1 macrophages are tumoricidal and highly microbicidal, and are often present after induction by interferon-γ (IFN-γ) in the tissue during infection or acute inflammation (Cook and Hagemann 2013). In contrast, alternatively activated M2 macrophages support T_H2-associated effector functions such as allergy in response to IL-4 or IL-13 and are involved with tissue remodeling and immunoregulation (Martinez and Gordon 2014). Tumor-associated macrophages (TAM) resemble M2 macrophages and have been implicated in fostering tumor progression, metastasis, angiogenesis, immune suppression of T-cell responses, and resistance to cancer treatment. The presence of TAM has been associated with decreased survival in breast, endometrial, prostate, ovary, gastric, head and neck, and oral carcinomas (Cook and Hagemann 2013; Zhang et al. 2012). Interestingly, improved overall survival was detected with TAM infiltration in colorectal cancer patients (Tan et al. 2005; Zhang et al. 2012). Emerging therapeutic strategies include the repolarization of TAM as a way of unlocking their anti-tumor potential. The effect of immunotherapy targeting CD40, a co-stimulatory molecule found on most APC, was examined to induce recognition and cytotoxicity of TAM by other cells of the immune system as an alternative to direct targeting (Beatty et al. 2011). An agonist CD40 antibody combined with gemcitabine chemotherapy activated macrophages that infiltrated the tumors and became tumoricial and facilitated the depletion of tumor stroma in a syngeneic pancreatic ductal adenocarcinoma mouse model. Macrophages in the tumors and spleens of these animals showed upregulation of MHC class II molecules, co-stimulatory factors including CD86, and cytokines such as IL-12, IFN-γ, and TNF-α. Together, these factors promote the 'M1'-like macrophage phenotype, improving the anti-tumoral adaptive immune response. In a similar way, since monocytes/macrophages express both FcεRs, tumor-specific IgE could be explored to 're-educate' TAM, leading to enhanced antigen presentation and targeting of tumor cells through ADCC and ADCP.

DC are unique APC because they are capable of stimulating naïve CD8 T lymphocytes via presentation of exogenous antigens in complex with MHC class I molecules, a process referred to as antigen cross-presentation (Janeway et al. 2005b; Spel et al. 2013). DC progenitors in the bone marrow give rise to circulating premature DC, which enter tissues such as the skin or mucosa as immature DC with high phagocytic capacity (Banchereau et al. 2000). After antigen capture, DC are activated by inflammatory cytokines. They then migrate to local lymphoid tissue and mature into highly specialized APC that can prime naïve T cells (Novak et al. 2010), expressing co-stimulatory molecules and inducing immune responses. DC, also present antigen in complex with MHC class II peptides to CD4 T-helper cells (Lemos et al. 2003), which secrete cytokines to recruit immune cells including antigen-specific CD8 CTL, as well as macrophages, eosinophils and NK cells (Banchereau et al. 2000), stimulating both primary and secondary T-cell responses (Maurer et al. 1996). The activated T cells help DC in terminal maturation, allowing lymphocyte expansion and differentiation. Importantly, effector cells are educated by DC to home to the site of tissue injury (Banchereau et al. 2000). FcεRI is expressed on DC (Kinet 1999), LC (Bieber et al. 1992; Osterhoff et al. 1994; Rieger et al. 1992), which are considered to be specialized DC present in the skin, and monocyte-derived DC (Novak et al. 2002). DC also express CD23 (Delespesse et al. 1991). The antigen uptake and presentation efficiency of DC are increased several fold in the presence of specific IgE (Daniels et al. 2010), leading to efficient activation of allergen-specific T cells. This suggests that tumor-targeted IgE can be employed to induce IgE-mediated tumor antigen presentation by DC, stimulating CD8 T cells, eventually resulting in lysis of tumor cells via CTL, and the establishment of long-term adaptive immune responses against the tumor.

Additional Advantages of IgE

Another property of IgE that could be advantageous for cancer therapy is its low endogenous blood levels, which provide less competition for receptor occupancy. IgE in circulation comprises only 0.02 % of total circulating immunoglobulin, compared to IgG at 85 % (Manz et al. 2005). In fact, high levels of endogenous IgG compete for host FcγR-expressing cells resulting in the loss of the therapuetic IgG from effector cells, thus high doses are required to overcome this limitation (Preithner et al. 2006). Importantly, IgE bound to effector cells has a long half-life from weeks to months (Achatz et al. 2010) compared to 2–3 days for IgG in tissues (Karagiannis et al. 2012), resulting in strong local retention of IgE by FcεRI-expressing resident immune effector cells even in the absence of antigens. The stable interaction of IgE and its receptors on effector cells enables efficient travel to the site of tumor, suggesting that tumor-specific IgE could be used to arm the immune system to target tumor cells. The local retention of IgE in tissues along with lower endogenous IgE levels in the blood may be translated to lower therapeutic doses and/or reduced frequency of administration compared to IgG therapy.

In contrast to IgG1, IgE does not elicit CDC (Janeway et al. 2005d). However, for many IgG-based immunotherapeutics, such as rituximab (Weng and Levy 2001) and trastuzumab (Hudis 2007; Peipp et al. 2008), ADCC is a major mechanism of action responsible for the antitumor activity of the antibody (Campoli et al. 2010; Weiner et al. 2012). CDC can be a hindrance to therapy as it has been shown that ADCC can be reduced by CDC interference (Wang et al. 2008, 2009). Thus, the fact that IgE does not mediate CDC should not necessarily dampen its potential effects as an anticancer agent. IgE can also mediate ADCC via activation of a variety of effector cells such as MC through the release of signaling molecules, enzymes, and cytokines as depicted in Fig. 2b. In addition, IgE mediates ADCP via activation of effector cells such as monocytes/macrophages normally abundant in tumors (Fig. 2b). Furthermore, there is no known inhibitory receptor for IgE; thus, its effector functions can be fully executed. IgG, on the other hand, can be bound by inhibitory FcγRIIb (CD32b), potentially decreasing ADCC/ADCP and antibody-mediated antigen presentation for IgG molecules (Clynes et al. 2000; Nimmerjahn and Ravetch 2007). These properties, together with all of the relevant characteristics of IgE mentioned above, make IgE an attractive strategy to target a wide range of malignancies.

2 IgE-based Immunotherapy

Below are several examples of IgE-based cancer therapies. The different strategies that have been evaluated in vivo are summarized in Table 1.

Table 1 Examples of IgE antibodies and mouse models used to evaluate their anti-tumor efficacy

IgE species	IgE specificity	Route of administration	Targeted cancer cells (route of cell inoculation)	Mouse model	References
Mouse	gp36 of MMTV	i.p.	H2712 mouse mammary carcinoma (s.c. and i.p.)	C3H/HeJ	Nagy et al. (1991)
Rat/human chimeric	Mouse Ly-2	s.c.	E3 mouse thymoma (s.c.)	C57BL/6	Kershaw et al. (1996)
Mouse	DNP	i.p.	MC38 mouse colon carcinoma cells expressing human CEA (s.c.)[a]	C57BL/6	Reali et al. (2001)
Mouse	DNP	s.c.	TS/A-LACK mouse mammary carcinoma cells coated with DNP (s.c.)	BALB/c	Nigro et al. (2009)
Mouse and mouse/human chimeric	Colorectal cancer antigen	i.v.	Human COLO 205 (s.c.)	SCID	Kershaw et al. (1998)
Rat/human chimeric	Mouse Ly-2	i.p.	E3 mouse thymoma (i.p.)	NOD–SCID	Teng et al. (2006)
Mouse/human chimeric	FBP	i.v.	IGROV-1 human ovarian carcinoma cells (s.c.)	C.B-17 scid/scid	Gould et al. (1999)
		i.p.	HUA patient-derived ovarian carcinoma (i.p.)	nu/nu	Karagiannis et al. (2003)
Mouse/human chimeric	NIP	s.c.	TS/A-LACK mouse mammary carcinoma cells coated with NIP (s.c.)	Human FcεRIα Tg BALB/c	Nigro et al. (2009)
Human	HER2/neu	i.p.	D2F2/E2 mouse mammary carcinoma cells expressing human HER2/neu (i.p.)	Human FcεRIα Tg BALB/c	Daniels et al. (2012a)
Human (truncated)	N/A	s.c.	TS/A-LACK mouse mammary carcinoma cells coated with truncated IgE (s.c.)	Human FcεRIα Tg BALB/c	Nigro et al. (2012)
Mouse/human chimeric	MUC1	s.c.	4T1 tumor cells expressing human MUC1 (s.c.)	Human FcεRIα Tg BALB/c	Teo et al. (2012)
Mouse/human chimeric	PSA	s.c.	CT26 tumor cells expressing human PSA (s.c.)	Human FcεRIα Tg BALB/c	Daniels-Wells et al. (2013)

[a] Tumor targeting occurred via a biotinylated anti-CEA IgG followed by streptavidin and then a biotinylated IgE.

s.c. subcutaneous, *i.p.* intraperitoneal, *i.v.* intravenous, *MMTV* mouse mammary tumor virus, *DNP* dinitrophenol, *FBP* folate binding protein, *NIP* nitrophenylacetyl, *HER2/neu* human EGFR2/neuroblastoma, *MUC1* mucin-1, cell surface associated, *N/A* not applicable, *PSA* prostate-specific antigen, *Tg* transgenic.

2.1 Passive Immunotherapy

2.1.1 Mouse IgE Specific for Glycoprotein 36 of the Mouse Mammary Tumor Virus

The use of IgE as an anticancer drug was pioneered by Nagy E. et al. in the 1990s (Nagy et al. 1991). Using the hybridoma technology, this group developed a mouse monoclonal IgE specific for the major envelope glycoprotein (gp36) of the mouse mammary tumor virus (MMTV) (Nagy et al. 1991). This anti-gp36 IgE demonstrated anti-tumor activity in C3H/HeJ mice challenged s.c. with a syngeneic MMTV-secreting mouse mammary adenocarcinoma (H2712). The mice were initially treated with 12.5 µg of IgE followed by the subsequent treatment of 25 µg of IgE via intraperitoneal (i.p.) injection every 4 days for 8 weeks. A 50 % protection rate was observed with IgE treatment compared to control animals. Similar protection was observed with mice injected i.p. with H2712 tumor cells treated with 25 µg of IgE every 4 days for 6 weeks. Long-term survival of IgE-treated mice was observed up to 175 days. This IgE did not protect mice challenged with MMTV-negative tumor cells (MA16/C). These results demonstrated, for the first time, the ability of a tumor-targeted IgE antibody to improve survival of tumor-bearing mice.

2.1.2 Rat/Human Chimeric IgE Specific for Murine Ly-2

The second tumor-specific IgE antibody developed was a rat/human chimeric IgE specific for mouse Ly-2, a subunit of the T-cell CD8 molecule that is also expressed on mouse T-cell tumors (Kershaw et al. 1996). Since human IgE does not interact with mouse FcεRI (Kinet 1999), a strategy utilizing a chimeric FcεRI was used to direct mouse CTL to recognize the tumor in a MHC-independent manner. This chimeric receptor was comprised of the human FcεRIα extracellular domain fused to the membrane proximal and transmembrane domains of human FcγRIIa and the cytoplasmic domain of human CD3ζ (Kershaw et al. 1996). In vitro, only mouse E3 thymoma cells (expressing Ly-2) pre-coated with the anti-(Ly-2) IgE were sensitive to lysis by mouse CTL (CTLL-R8) effector cells expressing the chimeric receptor (3H2). In an in vivo assay, 3H2 effector cells and E3 target cells at a 10:1 ratio in the presence or absence of the anti-(Ly-2) IgE were incubated together prior to s.c. injection into C57BL/6 mice. Although effector cells (3H2) alone significantly prolonged survival of mice, the antitumor effect was more prominent with addition of the anti-(Ly-2) IgE, where only 1 out of 5 mice developed tumors compared to 4 out of 5 mice with 3H2 cells alone. These results suggest that an anti-tumor IgE can be used to redirect engineered CTL to lyse tumor cells in an adoptive transfer therapy setting.

A similar strategy was used in immunodeficient non-obese diabetic–severe combined immunodeficiency (NOD–SCID) mice utilizing primary human T cells expressing a chimeric FcεRI (T-CL9) comprised of the extracellular domain of human FcεRIα, the human FcγRII hinge and transmembrane regions, the cytoplasmic signaling domain of human CD28, and the T-cell receptor ζ chain (Teng et al. 2006). T-CL9 treated with the anti-(Ly-2) IgE induced lysis of mouse E3 thymoma cells expressing Ly-2 in vitro. Production of immune stimulatory cytokines (IFN-γ and GM-CSF) was observed when the anti-(Ly-2) IgE was incubated with T-CL9 and E3 target cells. Importantly, increased survival from 18 to 27 days was observed in mice challenged i.p. with E3 tumors treated with T-CL9 cells (i.p.) loaded with the anti-(Ly-2) IgE at 6, 24, 48, and 96 h following tumor challenge compared with control mice. Similar enhanced survival was observed in mice with established tumors. However, complete tumor rejection and long-term survival were not observed, probably due to the fact that T-CL9 cells did not persist in the animal (Teng et al. 2006).

2.1.3 Mouse and Mouse/Human Chimeric IgE Specific for a Human Colorectal Cancer Antigen

A mouse IgE (mIgE 30.6) specific for a human colorectal cancer antigen was developed in parallel with a mouse/human chimeric IgE (chIgE 30.6) with the same variable regions (Kershaw et al. 1998). The variable regions of both antibodies were derived from the murine IgG2b antibody 30.6. Neither antibody showed direct in vitro cytotoxicity against human colorectal carcinoma COLO 205 cells. The growth of COLO 205 cells (s.c.) in SCID mice was significantly inhibited by 1 μg mIgE 30.6 injected intravenously (i.v.) 5 days after tumor challenge, although the effect was transient and lasted only for 48 h. This anti-tumor effect was not enhanced with 3 consecutive treatments of 1 μg mIgE 30.6 every 2 days starting 5 days post-tumor challenge. Administration of chIgE 30.6 did not result in anti-tumor activity, which is expected since human IgE does not bind murine FcεRs (Bettler et al. 1989; Kinet 1999). A previous study with a mouse/human chimeric IgG1 30.6 showed anticancer protection with 250 μg of antibody administered at days 5, 7, and 9 after tumor challenge (Mount et al. 1994). The strong anticancer activity exhibited by the murine IgE may be due to its higher affinity for its FcεR compared to the affinity of IgG to its FcγR. It is notable that chIgG1 30.6 and chIgE 30.6 were not tested simultaneously. It was suggested that the release of cytokines, such as TNF-α or IL-4, from MC may be involved in the anti-tumor effect of mIgE 30.6 (Kershaw et al. 1998).

2.1.4 Mouse/Human Chimeric IgE Specific for Human FBP

Folate binding protein (FBP), also known as folate receptor α, is overexpressed on about 80 % of serous non-mucinous ovarian adenocarcinomas (Veggian et al. 1989). A mouse/human chimeric IgE specific for FBP (MOv18-IgE) was developed and has been extensively studied (Gould et al. 1999). The anti-tumor efficacy of this IgE was compared to a mouse/human chimeric IgG1 containing the same variable

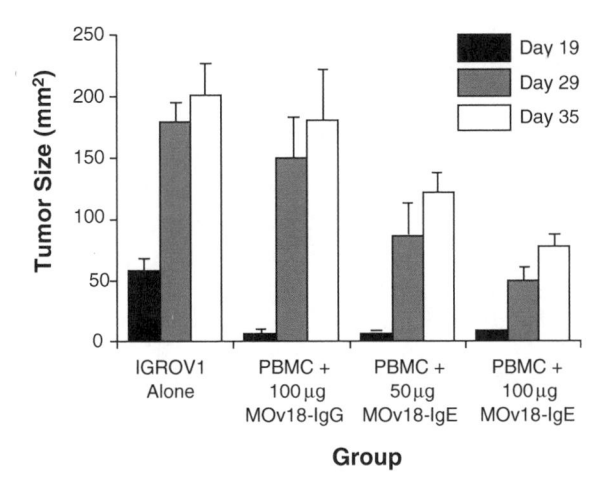

Fig. 3 MOv18-IgE inhibits IGROV1 human ovarian cancer cell growth in vivo in the presence of human PBMC effector cells. A matched variable region set of mouse/human chimeric IgE and IgG1 antibodies specific for FBP was tested for their ability to inhibit tumor growth in a SCID transgenic mouse model of ovarian cancer. SCID mice were pre-treated with rabbit anti-asialo G_{M1} antibody i.p. to eliminate host NK cells. One day later, all mice were challenged s.c. with 2.5×10^6 IGROV1 carcinoma cells, followed by i.v. administration of 3×10^6 human PBMC combined with either 100 μg MOv18-IgG1, 100 μg MOv18-IgE, or 50 μg MOv18-IgE. Mean tumor size ± standard error of the mean measured on days 19, 29, and 35 after challenge ($n = 4$ for all groups) is shown. Mice given PBMC and 50 μg MOv18-IgE exhibited reduced growth by both day 29 (51 % inhibition; $p < 0.03$) and day 35 (40 % inhibition; $p < 0.04$). Mice injected with 100 μg MOv18-IgE exhibited an even greater degree of inhibition, relative to the positive tumor control, on day 29 (72 % inhibition; $p < 0.001$) and day 35 (62 % inhibition; $p < 0.005$). Adapted from Fig. 4 of Gould et al. (1999) with kind permission from John Wiley and Sons

regions in 2 different human xenograft transgenic models of FBP-expressing human ovarian carcinoma grown in immunodeficient mice. Mice were challenged s.c. with human ovarian carcinoma IGROV1 cells followed by i.v inoculation of human peripheral blood mononuclear cells (PBMC) as a source of effector cells capable of binding human IgE, in the presence of MOv18-IgE or MOv18-IgG1 (100 μg). MOv18-IgE demonstrated a superior and prolonged anti-tumor effect up to 35 days compared to MOv18-IgG1-treated mice, which developed tumors after 19 days (Fig. 3). Of note, treatment with half the dose of IgE (50 μg) showed 40 % inhibition of tumor growth at day 35. Since the human Fc region of the chimeric IgE does not bind mouse FcεRs (Bettler et al. 1989; Kinet 1999), MOv18-IgE did not induce an anti-tumor response in the absence of human PBMC. This IgE-mediated tumor response is expected to be amplified in humans with a permanent supply of the entire effector cell repertoire, including MC, basophils, monocytes/macrophages, eosinophils, as well as an intact immune system. A similar model employed nude mice challenged i.p. with human HUA ovarian cancer cells expressing FBP inoculated i.p. with human PBMC alone or in combination with the MOv18-IgE or Mov18-IgG1 (Karagiannis et al. 2003). MOv18-IgE significantly increased survival up to

40 days compared to 22 days with MOv18-IgG1. Immunohistochemistry analysis of HUA xenograftsrevealed infiltration of human monocytes into tumor lesions, reinforcing their role in the anti-tumor efficacy of tumor-specific IgE, which was confirmed using monocyte-depleted PBMC in this tumor model (Karagiannis et al. 2007). Subsequent flow cytometric analysis revealed the distinct pathways by which human monocytes and eosinophils mediated MOv18-IgE-dependent tumor killing in vitro (Karagiannis et al. 2007), where IGROV1 cells were primarily lysed by antibody-dependent cell-mediated cytotoxicity (ADCC) via binding to FcεRI, and phagocytosed by ADCP via binding to CD23, as depicted in Fig. 2.

The safety of the MOv18-IgE has been evaluated in ex vivo studies using blood from patients with ovarian cancer. The potential FcεRI cross-linking capacity of soluble Folate binding protein (FBP), which is found in monomeric form in the circulation (Karagiannis et al. 2012), was recently assessed (Rudman et al. 2011). Even though high levels of Folate binding protein (FBP) can be found in the blood of cancer patients, concentrations up to 10-fold higher than those observed in the blood of patients did not induce degranulation of RBL SX-38 (rat basophil leukemic cells expressing human FcεRI) in the presence of the MOv18-IgE. The ability of cancer cells to induce degranultion was also explored. IGROV-1 human ovarian cancer cells expressing Folate binding protein (FBP) were incubated with the MOv18-IgE and RBL SX-38 effector cells. Degranulation was induced at high tumor:effector cell ratios, as is expected at tumor sites. Importantly, degranulation was not observed at lower ratios that coincide with numbers of cancer cells that may occur in the peripheral blood of ovarian cancer patients. Incubation with tumor cells not expressing Folate binding protein (FBP) did not induce degranulation, confirming the need of a tumor-targeted IgE. basophils, either isolated or within the whole blood from patients, were also not activated by MOv18-IgE in the presence of soluble Folate binding protein (FBP), as evaluated via CD63 upregulation. Even though antibody-induced tumor cell death could release fragments of tumor cells from lesions that might activate basophils in the periphery, previous studies suggest that in addition to cytotoxicity, MOv18-IgE triggers substantial infiltration and phagocytosis by monocytes/macrophages (Karagiannis et al. 2007, 2008), which may limit their escape into the circulation. In addition, although IgG autoantibodies against FBP were detected in low concentrations in a small subset of patients, a hypersensitivity reaction was not induced in vitro (Rudman et al. 2011). These studies show that the extent of Folate binding protein (FBP) overexpression on tumor cells and the density of tumor cells seem to determine IgE/FcεR cross-linking capacity, with the expectation of activating FcεRI-mediated effector cells within the tumor mass but not in the circulation.

2.1.5 Engineered IgE Specific for Human HER2/*neu*

The anti-tumor efficacy of IgE antibodies specific for HER2/*neu* (human EGFR2/ neuroblastoma, ErbB2), which is overexpressed by approximately 20–30 % of all breast and ovarian cancers and is associated with poor prognosis (Berchuck et al. 1990;

Meden et al. 1994; Slamon et al. 1987, 1989), has also been evaluated. HER2/*neu* is a member of the EGFR family and has tyrosine kinase activity that mediates cell survival and proliferation (Martinelli et al. 2009). Trastuzumab (Herceptin®), a humanized IgG1, has shown efficacy as a treatment for metastatic breast cancer. However, the majority of patients treated with trastuzumab alone or combined with chemotherapeutic agents eventually relapse (Ahn and Vogel 2011). In addition, a significant number of breast cancer patients with HER2/*neu* expression do not respond to trastuzumab-based therapies (Ahn and Vogel 2011; Fernandez et al. 2010; Nahta et al. 2009), confirming the need for additional strategies in targeting HER2/*neu* overexpressing tumors. Thus, the variable regions of trastuzumab were fused to the constant region of human IgE to produce a humanized trastuzumab IgE (Karagiannis et al. 2009). Similar direct cytotoxic effects up to 48 h were observed with both antibodies in cell viability assays using SK-BR-3 human breast cancer cells. Trastuzumab IgE mediated antibody-dependent cell-mediated cytotoxicity (ADCC) in the presence of U-937 human monocytic cells, while trastuzumab IgG1 exhibited a different mechanism, , as detected by flow cytometry and validated by confocal microscopy. Degranulation was induced by trastuzumab IgE in the presence of murine colon carcinoma cells expressing human HER2/*neu* (CT26-HER2/*neu*) incubated with rat basophilic leukemia cells that express human FcεRI (RBL SX-38).

We have developed a fully human anti-HER2/*neu* IgE using the variable regions of the single-chain Fv C6MH3-B1 (Daniels et al. 2012a). C6MH3-B1 was isolated from a naïve human phage display library, affinity matured in vitro, and does not compete with trastuzumab for HER2/*neu* binding (Schier et al. 1995, 1996; Tang et al. 2007). This IgE induced in vitro degranulation of RBL SX-38 cells expressing human FcεRI in the presence of murine mammary carcinoma cells expressing human HER2/*neu* (D2F2/E2) but not in the presence of the parental D2F2 cells that lack HER2/*neu* expression or shed (soluble) extracellular domain of HER2/*neu* (ECD^{HER2}) (Daniels et al. 2012a). These results suggest that the acute inflammatory response (type I hypersensitivity) is expected to occur within the tumor microenvironment, where the HER2/*neu* antigen is overexpressed at high levels on the surface of cancer cells (Pegram and Ngo 2006), facilitating FcεRI cross-linking and triggering effector cell degranulation. However, due to the mono-epitopic nature of the interaction between the C6MH3-B1 IgE and ECD^{HER2}, such a reaction is not expected to occur in the presence of this soluble antigen. This is supported by our data in vivo (Daniels et al. 2012a), which show that ECD^{HER2} did not induce a reaction in the presence of the IgE. Administration of the C6MH3-B1 IgE intradermally (i.d.) into mice, followed by i.v. injection of an anti-human κ antibody to artificially cross-link the IgE-loaded FcεRI, mediated a local, passive cutaneous anaphylaxis reaction, as observed via leakage of Evans blue dye into the skin (Daniels et al. 2012a). This anaphylactic response was not observed with systemic administration of ECD^{HER2}, confirming that soluble ECD^{HER2} in the presence of the C6MH3-B1 IgE cannot induce cross-linking of the FcεRI. In addition, priming of human DC with complexes of ECD^{HER2} and the C6MH3-B1 IgE resulted in CD4 and CD8 T-cell activation in vitro (Daniels et al. 2012a), consistent with the possibility of a tumor-specific IgE enhancing DC antigen presentation via MHC class I

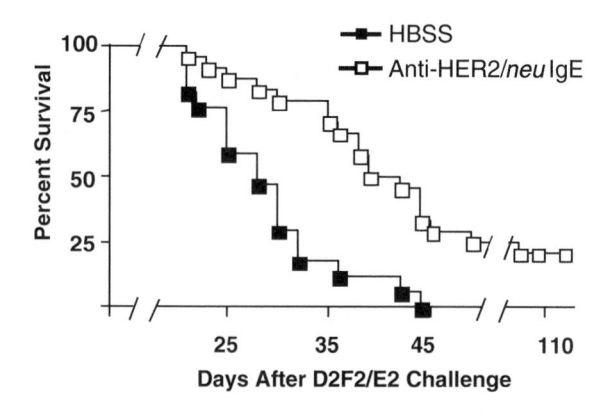

	# of animals	Median survival
HBSS	17	28
Anti-HER2/*neu* IgE	18	39

Fig. 4 C6MH3-B1 IgE prolongs survival of human FcɛRIα transgenic mice bearing syngeneic mammary tumors expressing human HER2/*neu*. Survival plot of female BALB/c human FcɛRIα transgenic mice challenged i.p. with a lethal dose of D2F2/E2 (2×10^5 cells) mammary cancer cells on day 0. Animals were treated i.p. with 100 μg C6MH3-B1 IgE or buffer alone on days 2 and 4. Data shown are the combined results from 3 independent experiments. The experimental group ($n = 18$) compared to buffer control ($n = 17$) shows a significant improvement in survival ($p < 0.001$, log-rank test). The remaining 5 long-term survivors were considered cured. Adapted from Fig. 6 of Daniels et al. (2012a) with kind permission from Springer Science and Business Media

and II pathways. Lower levels of T-cell activation were observed with complexes of ECD^{HER2} and trastuzumab in the same assay. This suggests that the C6MH3-B1 IgE can bind soluble ECD^{HER2} and redirect it to APC enhancing antigen presentation. Complexes of the C6MH3-B1 IgE-ECD^{HER2} alone or the use of DC loaded with these complexes could potentially be used as a tumor vaccine.

The C6MH3-B1 IgE was well-tolerated and significantly prolonged survival of BALB/c human FcɛRIα transgenic mice bearing a mammary tumor expressing human HER2/*neu* (D2F2/E2), when injected i.p. in a model of peritoneal carcinomatosis (Fig. 4). The human FcɛRIα transgenic mouse (also available in C57BL/6 background) has human FcɛRIα knocked in and murine FcɛRIα knocked out (Dombrowicz et al. 1996, 1998; Fung-Leung et al. 1996) and is an important model given that human IgE does not interact with mouse FcɛRI (Kinet 1999). In this mouse model, the human FcɛRIα has been shown to assemble with endogenous mouse FcɛRβ and FcɛRIγ to form functional receptors that mediated anaphylactic reactions when stimulated by human IgE (Dombrowicz et al. 1998). In addition to MC and basophils, in this mouse model, human FcɛRIα is also expressed on murine monocytes/macrophages, DC, LC, and eosinophils (Dombrowicz et al. 1998; Fung-Leung et al. 1996; Kinet 1999), similar to the pattern of expression observed in humans. The C6MH3-B1 IgE was also well-tolerated in an initial study in non-human primates (Daniels et al. 2012a). Human IgE has been shown to mediate anaphylaxis in cynomolgus monkeys (*Macaca fascicularis*) (Weichman et al. 1982), which has been used for target validation and assessment of therapeutic candidates designed to interfere with IgE biology, due to the highly conserved nature of FcεRIα between cynomolgus monkey and humans, with a sequence similarity of 93.8 % (Saul et al. 2014). The monkeys have also been used for safety studies on other HER2/*neu*-targeted antibody therapies, including trastuzumab (Junutula et al. 2010), due to the high sequence homology between human and cynomolgus ECD^{HER2} (Adams et al. 2006). In the initial study on the safety of the C6MH3-B1 IgE, animals received i.v. infusion of either a 0.0024 or 0.08 mg/kg dose (10 ml over 20 min) (Daniels et al. 2012a). These doses were chosen based on previous studies in humans using different immune stimulating therapeutic antibodies and their derivatives (Braly et al. 2009; King et al. 2004). No changes in eating habits or general health were observed for one week after dosing. This preliminary study demonstrated, for the first time, that systemic infusion of an IgE targeting a tumor antigen is and well-tolerated in non-human primates, suggesting a manageable safety profile allowing the anti-HER2/*neu* IgE to mobilize multiple anticancer immune mechanisms.

2.1.6 Mouse/Human Chimeric IgE Specific for Human CD20

CD20 is a transmembrane protein expressed on B cells during early pre-(B-cell) development until its downregulation when cells differentiate into antibody-secreting plasma cells (Reff et al. 1994). CD20 is expressed at reasonably high levels, is not internalized after antibody binding (Liu et al. 1987), and is not shed or secreted into the circulation, allowing efficient targeting of malignant B cells (Grillo-Lopez 2000). Even though the mouse/human chimeric anti-CD20 IgG1 rituximab (Rituxan®) is currently used in clinic for the treatment of NHL, resistance to this antibody is a common occurrence (Rezvani and Maloney 2011; Smith 2003), indicating the need for additional treatment strategies.

A mouse/human chimeric anti-human CD20 IgE has been developed (Teo et al. 2012). This IgE incubated with cord blood-derived MC and OCI-Ly8 human B lymphoma target cells showed an increase in tumor cytotoxicity 24 h after treatment compared to a non-targeted IgE (Fig. 5). Interestingly, this IgE-mediated cytotoxicity was decreased by the addition of an anti-TNF-α antibody, confirming its involvement in MC-mediated cell death. Similarly, antibody-dependent cell-mediated cytotoxicity (ADCC) was observed with eosinophils differentiated from cord blood mononuclear cells and the IgE (Fig. 5). The addition of heparin blocked tumor cell death, probably by neutralizing cationic proteins released by eosinophils. This study shows that MC and eosinophils differentiated in vitro can be used as effector cells in IgE-mediated antibody-dependent cell-mediated cytotoxicity (ADCC) assays.

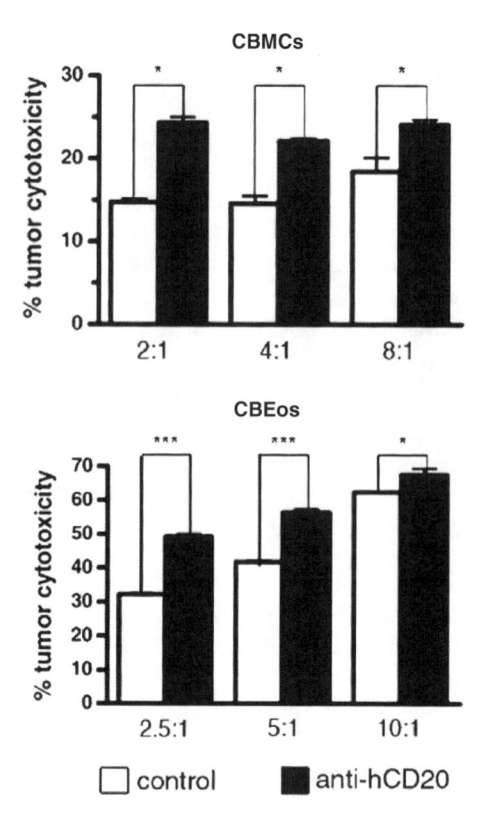

Fig. 5 An anti-CD20 IgE induces ADCC in vitro in the presence of MC and eosinophils. OCI-Ly8 human B cell lymphoma cells labeled with CFSE were incubated with unstained CBMCs (cord blood-derived MC) and 2.5 µg/ml anti-CD20-IgE, or cord blood-derived eosinophils (CBEos) and 5 µg/ml anti-CD20-IgE at different effector:target ratios. The cell mixture was stained with propidium iodide (PI) before flow cytometry analysis. The percentage of PI$^+$ cells in the CFSE^{hi} fraction representing total tumor cytotoxicity is shown. Results shown are mean ± standard deviation (SD) of one representative experiment. Student's t-test *, $p < 0.05$; ***, $p < 0.005$. Adapted from Fig. 3 of Teo et al. (2012) with kind permission from Springer Science and Business Media

2.1.7 Mouse/Human Chimeric IgE Specific for Human MUC1

MUC1 (Mucin 1, cell surface associated) is expressed at the apical surface of healthy epithelial cells and is characterized by a high glycosylation level (Roulois et al. 2013). It is overexpressed on tumors arising from glandular epithelium, such as the breast, ovary, pancreas, and colon cancers (Gourevitch et al. 1995; Kotera et al. 1994), with a loss of polarity and modification of its glycosylation pattern (Roulois et al. 2013). A mouse/human chimeric IgE specific for the human MUC1 antigen has been developed (Teo et al. 2012). The efficacy of the anti-MUC1 IgE was tested in vivo using a murine breast carcinoma cell line expressing the transmembrane form of human MUC1 (4T1.hMUC1). A 24 % reduction in tumor size was observed in human FcεRIα transgenic mice challenged s.c. with 4T1. hMUC1 cells (10^5) and treated with 20 μg anti-MUC1 IgE on days 1 through 5. While 3 of 8 mice in the control group survived up to 34 days, 5 of 8 mice in the anti-MUC1 IgE-treated group were still alive at this time. Mice treated with the anti-MUC1 IgE showed an increased presence of MC in necrotic peritumoral regions exhibiting signs of degranulation, compared to mice treated with a control antibody. The lack of a stronger response was attributed to the fact that 4T1 tumors are highly avascular and grow in a densely packed mass, which may impede drug delivery or effector cell recruitment. An alternative approach was implemented to further evaluate the anti-tumor activity of the anti-MUC1 IgE. 4T1 cells were engineered to express mouse IgE specific for human MUC1 to ensure continual exposure of the tumor to the antibody. In addition, 4T1.hMUC1 cells were engineered to express IL-5, which has been shown to be involved in recruiting eosinophils into the peritumoral stroma (Sanderson 1988) or monocyte chemoattractant protein-1 (MCP-1) to recruit monocytes and MC (Soucek et al. 2007), thus increasing the frequency of key effector cells at the tumor site. The 3 engineered tumor cell lines were inoculated s.c. into human FcεRIα transgenic mice. Rejection of tumors was observed up to 40 days in 7 of 8 mice, suggesting a durable response. Tumor elimination was not observed in the absence of cells expressing the anti-MUC1 IgE. Interestingly, rejection of tumors up to 30 days was also observed with parental 4T1 cells inoculated s.c. into the opposite flank of the 7 mice that rejected the initial tumor challenge, suggesting these animals developed memory responses and that these responses targeted other tumor antigens, indicating that epitope spreading had occurred. These results also suggest that a combination of antigen-specific IgE and chemokine-transfected tumor cells may possibly be used as a tumor vaccine.

2.1.8 Human IgE Specific for Human EGFR

EGFR is overexpressed in a variety of tumors including breast, ovary, prostate, colon, gastric, liver, pancreas, lung, brain, and bladder cancer (Salomon et al. 1995). Two IgE antibodies with human contant regions containing the variable regions of cetuximab, a mouse/human chimeric IgG1 (Erbitux®), and matuzumab (EMD 72000), a humanized IgG1, were developed (Spillner et al. 2012), targeting spatially distinct structures on EGFR (Schmiedel et al. 2008). Proliferation and cytotoxicity assays demonstrated inhibition of EGFR signaling and direct tumor cytotoxicity of both IgE antibodies (Spillner et al. 2012). Interestingly, using purified human monocytes as effector cells, the cetuximab IgE increased ADCC up to 70 % compared to 30 % by its IgG counterpart using A431 human epidermoid carcinoma target cells, while ADCP levels were similar. This study also demonstrated that surface clustering of the EGFR, and not soluble EGFR, triggers IgE-mediated degranulation.

2.1.9 Mouse/Human Chimeric IgE Specific for Human PSA

The high rate of relapse and absence of a cure for metastatic prostate cancer, combined with the presence of organ-specific tumor antigens and the initial slow-growing nature of the disease, make prostate cancer an attractive target for immunotherapy. The prostate-specific antigen (PSA) is a secreted molecule that has been routinely used for blood screening of prostate cancer and assists in assessing responses to therapy and determining tumor progression (Obort et al. 2013). The anti-PSA murine monoclonal IgG1 AR47.47 complexed with PSA showed enhanced antigen presentation by human DC and induced both CD4 and CD8 T-cell activation (Berlyn et al. 2001). To examine if an IgE also possesses this activity, our group constructed a novel mouse/human chimeric anti-PSA IgE composed of the variable regions of AR47.47 (Daniels-Wells et al. 2013). An artificial multi-epitopic molecule termed human serum albumin (HSA)-PSA, which consists of multiple PSA peptides containing the AR47.47 binding epitope conjugated to HSA, induced degranulation of rat basophil cells expressing human FcεRI (RBL SX-38), confirming the functionality of this IgE. In contrast, incubation of the anti-PSA IgE with free (soluble) PSA or the naturally occurring mono-epitopic PSA-ACT (α1-antichymotrypsin) complex with RBL SX-38 did not induce degranulation, even in amounts exceeding those detected in the blood of cancer patients (Fig. 6a). Degranulation mediated by the anti-PSA IgE was also examined in vivo using BALB/c human FcεRIα transgenic mice (Daniels-Wells et al. 2013). Administration of the anti-PSA IgE (i.d.) into mice, followed by i.v. injection of an anti-human κ antibody to artificially cross-link IgE-loaded FcεRI, mediated a local, passive cutaneous anaphylaxis reaction, as observed via leakage of Evans blue dye into the skin (Fig. 6b). However, this anaphylactic response was not observed with systemic administration of PSA at concentrations exceeding those found in the blood of cancer patients in the same model, confirming that free PSA in the presence of the IgE does not induce cross-linking of the FcεRI. The observation of degranulation when artificially cross-linked suggests the ability of the IgE to induce degranulation in the tumor microenvironment, where cross-linking of the FcεRI may occur due to a high PSA density in the tumors. In addition, priming of human DC with complexes of PSA and the anti-PSA IgE resulted in CD4 and CD8 T-cell activation in vitro. This suggests the possibility of the anti-PSA IgE to complex with PSA in the blood of patients leading to the induction of a secondary immune response involving CTL activity. Additionally, anti-PSA antibodies have been previously used for therapeutic purposes (Katzenwadel et al. 2000) and for the delivery of chemotherapeutic drugs (Sinha et al. 1996, 1999), inhibiting the growth of PSA-producing tumors. Furthermore, a radiolabeled anti-PSA has been successfully used to image PSA-producing tumor cells (Evans-Axelsson et al. 2012). These studies on PSA-targeted antibodies suggest that the anti-PSA IgE may also be able to target the tumor directly due to the high PSA density within the tumor microenvironment.

An initial prophylactic vaccination study was conducted using PSA complexed with the anti-PSA IgE in human FcεRIα transgenic male BALB/c mice (Daniels-Wells et al. 2013). The vaccination was well-tolerated and significantly prolonged

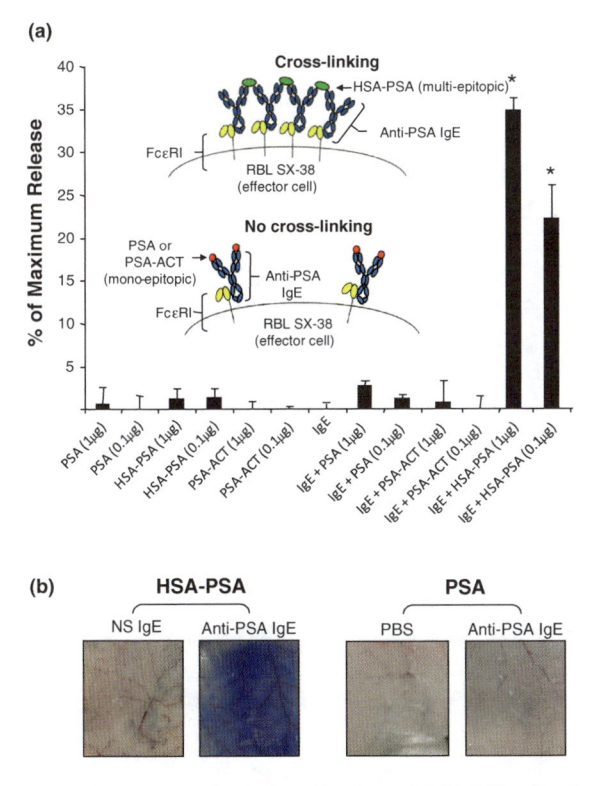

Fig. 6 In vitro and in vivo degranulation induced by the anti-PSA IgE. **a** Rat basophil leukemic cells expressing human FcεRI (RBL SX-38) were sensitized with 1 μg of IgE or buffer alone for 2 h, followed by a 2-h incubation with PSA, PSA-ACT, or HSA-PSA (an artificial PSA multi-epitope conjugate). HSA-PSA is a positive control of a multi-epitopic antigen. Antigen concentration was adjusted to the equivalent molar amount of PSA-ACT (1 or 0.1 μg/ml). The release of β-hexosaminidase from cells was monitored through the addition of 2.5 μM p-nitrophenyl-N-acetyl-β-D-glucosamine in 50 mM citrate buffer (pH 4.5). Mean ± SD of triplicate samples are shown. Data are representative of 2 independent experiments ($p < 0.01$, Student's t-test compared to either component alone). **b** In vivo local, passive anaphylaxis in human FcεRIα transgenic mice. Mice were injected i.d. on the back with buffer (PBS) alone, 1 μg of PSA IgE, or a non-PSA-specific IgE (NS IgE). Four hours later, mice were injected i.v. with either 50 μg of free PSA or 50 μg of the HSA-PSA peptide conjugate in 1 % Evans blue in PBS. Animals were euthanized 10 min later. Local hypersensitivity (anaphylactic) responses induced by degranulation of FcεRI-expressing cells in the skin were visualized by leakage of the dye into the skin due to vasodilation of local blood vessels. A reaction was detected with anti-PSA IgE and HSA-PSA (artificial multi-epitope) but not with anti-PSA IgE and PSA. Representative images of 3 independent experiments are shown. Adapted from Figs. 2 and 3 of Daniels-Wells et al. (2013) with kind permission from BioMed Central

the survival ($p \leq 0.01$ compared to the other groups, log-rank test) of mice challenged with syngeneic human PSA expressing murine colon carcinoma tumors (CT26-PSA) (Fig. 7). No protection was observed with an anti-PSA IgG1 complexed with PSA or with PSA alone. Importantly, this protective IgE-mediated

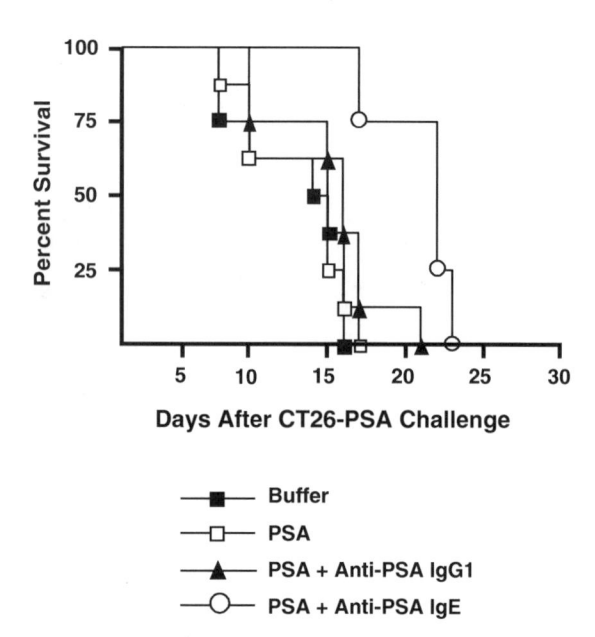

Fig. 7 Vaccination study with the anti-PSA IgE in human FcεRIα transgenic mice. Survival plot of human FcεRIα transgenic mice (male) vaccinated s.c. with PSA alone (4 μg) or complexed in a 1:1 molar ratio to either the anti-PSA IgE or IgG1 in the left flank (day 0). Mice were given a booster on day 15 in the same left flank. On day 40, mice were challenged with 10^6 mouse colon carcinoma CT26 cells expressing human PSA (CT26-PSA) s.c. in the right flank. Animals were euthanized when tumors reached 1.5 cm in diameter. Animals administered with buffer alone were used as negative controls. Combined data from 2 independent experiments with a total of 8 mice per group are shown. Reprinted from Fig. 5 of Daniels-Wells et al. (2013) with kind permission from BioMed Central

response was associated with significant increased levels of mouse anti-PSA IgG2a, an isotype connected to the T helper 1 (T_H1) immune response in mice (Finkelman et al. 1990). Even though the in vivo effects observed with the anti-PSA IgE under these initial conditions were modest (although statistically significant) (Daniels-Wells et al. 2013), the observed anticancer effect could have major clinical significance since the BALB/c strain used in these studies is biased toward a T_H2 response (Chu et al. 1997). In addition, these results also suggest the potential use of the anti-PSA IgE complexed to PSA or the use of DC loaded with these complexes as a vaccination approach.

2.2 Active Immunotherapy (Vaccination Approaches)

The above applications describe examples of passive immunotherapy, which in certain cases may also be used as vaccine approaches. Below, we will focus on exclusive vaccination approaches involving IgE.

2.2.1 Oral Mimotope Vaccination

An alternative strategy to the passive administration of an IgE is to induce an endogenous IgE response. A novel approach to establish an active vaccination protocol is to induce tumor antigen-specific IgE through the oral route (Riemer et al. 2007). Synthetically manufactured epitope mimics (mimotopes) were generated for the epitope of human HER2/*neu* that is recognized by trastuzumab. These mimotopes were used to immunize immunocompetent BALB/c mice orally under simultaneous neutralization and suppression of gastric acid, a feeding regimen shown to effectively induce T_H2 immune responses (Scholl et al. 2005b; Untersmayr et al. 2003). The induction of high-titer serum IgE targeting the HER2/*neu* antigen was observed (Riemer et al. 2007). In addition, the endogenous anti-HER2/*neu* IgE recognized HER2/*neu*-expressing human breast cancer cells (SK-BR-3), resulting in both degranulation and cytotoxicity of rat basophil cells expressing rodent FcεRI (RBL-2H3).

2.2.2 IgE-coated Cellular Approaches

The adjuvant effect of IgE and its role in the design of new cell-based tumor vaccines have been explored. Targeting of tumors by IgE was performed using a three-step strategy based on biotin–avidin interactions (Reali et al. 2001), which has shown high efficiency in targeting cytotoxic molecules on tumor cells in vivo (Guttinger et al. 2000). First, tumor cells were coated with a biotinylated IgG antibody targeting the tumor antigen. Next, streptavidin was added to bind biotin-labeled tumor cells, followed by binding of a biotinylated IgE to the streptavidin–tumor complex to initiate effector functions (Reali et al. 2001). Immunocompetent C57BL/6 mice were inoculated s.c. with murine colon adenocarcinoma tumors expressing human carcinoembryonic antigen (MC38-CEA-2) and treated with biotinylated mouse anti-CEA IgG2b (i.p.) 2 days later, followed by avidin (i.p.) on day 3 to eliminate the unbound IgG antibodies in the blood, and streptavidin (i.p.) 4 h later to create the biotin–avidin bridge with the bound anti-CEA IgG2b antibody. On day 4, biotinylated murine anti-dinitrophenyl (DNP) IgE or IgG control antibody was administered (i.p.). IgE treatment significantly reduced tumor growth and prolonged survival compared with IgG treatment. Two out of 10 mice receiving IgE treatment completely rejected the tumor and resisted subsequent tumor challenge with the parental MC38 cell line, consistent with the generation of a memory response and epitope spreading. Depletion of eosinophils, as well as CD4 and CD8 T cells, abrogated the anti-tumor effects, suggesting the role of these cells in IgE-mediated responses against tumors. These findings were confirmed using the highly aggressive syngeneic murine lymphoma Rauscher virus-induced T-cell lymphoma cells (RMA) expressing tumor antigen Thy 1.1 under similar conditions (Reali et al. 2001).

Cellular vaccines were developed using irradiated tumor cells coated with mouse IgE or IgG by the biotin–avidin bridging technique described above (Reali et al. 2001). Immunocompetent C57BL/6 mice were s.c. vaccinated twice (2 weeks apart) with IgE- or IgG-loaded irradiated (10,000 rad) MC38-CEA-2 cells. Two weeks after the second immunization, animals were challenged with parental MC38 tumor cells. All doses of cells coated with IgE showed significant delay in tumor growth, while only the highest dose of cells (30×10^4) coated with IgG showed protection from tumor challenge. Similar results were observed using a similar vaccination schedule in the RMA tumor model (Reali et al. 2001). These studies show the efficacy of using IgE to stimulate T cell-mediated immune responses, including epitope spreading and induction of a memory response, confirming its potential role as an adjuvant of anti-tumor vaccines.

A slightly different strategy was used to confirm the adjuvant effect of IgE-coated tumor cells. TS/A-LACK murine mammary adenocarcinoma cells were infected with modified vaccinia virus Ankara (MVA) in a BALB/c model, avoiding the need for irradiation of tumor cells prior to vaccination and increasing the tumorigenicity of infected tumor cells (Nigro et al. 2009). TS/A-LACK cells were subsequently haptenized with DNP then coated with a mouse anti-DNP IgE. Immunocompetent mice were vaccinated s.c. and challenged 15 days later with live TS/A-LACK cells. Anti-tumor protection was achieved after a single immunization with IgE-loaded cells and was comparable to mice vaccinated twice in the absence of IgE. FcεRI was demonstrated to be the mediator of IgE activity in this anti-tumor vaccination study. To translate these findings into a human system, the same experiments were performed using the human FcεRIα transgenic mouse with TS/A-LACK cells haptenized with NIP (nitrophenylacetyl) and loaded with a mouse/human chimeric anti-NIP IgE (Nigro et al. 2009). Significant anti-tumor protection was observed in human FcεRIα transgenic mice immunized with IgE-loaded tumor cells (Fig. 8a), but not in wild-type BALB/c mice (Fig. 8b) since human IgE does not interact with mouse FcεRI (Kinet 1999). Taken together, these studies suggest that IgE can act as an adjuvant of immunization in cancer therapy.

Based on these results, this group developed a novel protocol employing membrane IgE for anti-tumor vaccination to eliminate any possible anaphylactogenicity caused by circulating IgE (Nigro et al. 2012). Truncated human IgE lacking the Fab regions (tmIgE) engineered into a recombinant MVA (rMVA-tmIgE) was used to infect TS/A-LACK tumor cells, resulting in transport of tmIgE to the infected cell surface. Human FcεRIα transgenic mice were vaccinated s.c. and challenged 15 days later with live TS/A-LACK cells. Mice immunized with rMVA-tmIgE-infected TS/A-LACK cells showed significant attenuation of tumor

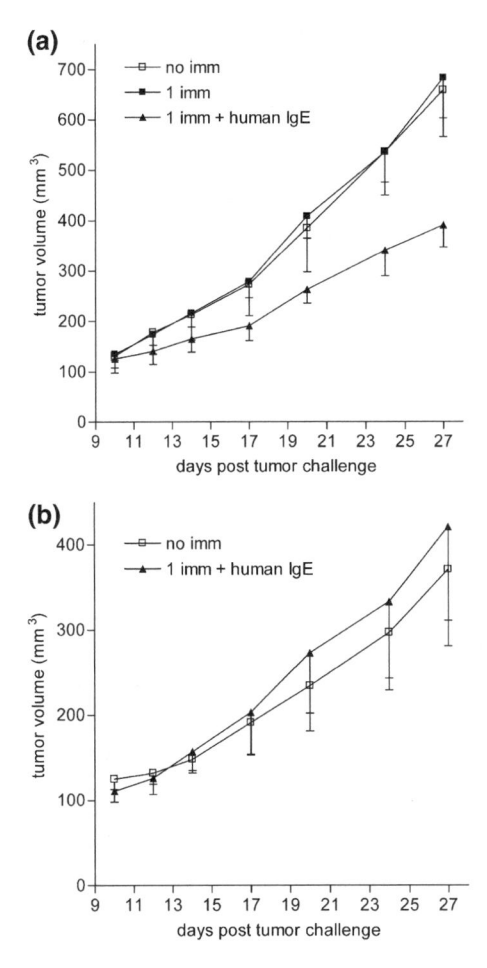

Fig. 8 IgE adjuvant effect in human FcεRIα transgenic mice. **a** Mice were s.c. vaccinated with 10^5 mouse/human chimeric anti-NIP IgE-loaded (1 imm + human IgE) or IgE-free (1 imm) MVA-infected TS/A-LACK cells at day -15 ($p < 0.0001$). At day 0, all vaccinated mice were challenged s.c. with 2×10^5 living TS/A-LACK cells. Non-immunized mice challenged with living TS/A-LACK cells were used as controls (no imm). **b** As a control, wild-type BALB/c mice were s.c. vaccinated with 10^5 human IgE-loaded (1 imm + human IgE) MVA-infected TS/A-LACK cells at day -15. At day 0, all vaccinated mice were challenged s.c. with 2×10^5 living TS/A-LACK cells. Non-immunized mice challenged with living TS/A-LACK cells were used as controls (no imm). Results are representative of 3 experiments. Mean ± standard error of the means is shown. Reprinted from Fig. 5 of Nigro et al. (2009) with kind permission from The American Association of Immunologists

growth compared with mice immunized with control vaccine not expressing tmIgE. This effect is comparable to that obtained using secretory IgE and was lost in wild-type BALB/c mice, confirming the role of FcεRI in IgE-active anti-tumor immunotherapy (Nigro et al. 2012).

3 Safety Concerns

A major perceived concern in using IgE in the passive immunotherapy of cancer is the potential to induce a systemic type I hypersensitivity reaction (anaphylactic shock). However, there is evidence suggesting that this adverse reaction may not occur. Type I hypersensitivity reactions are only induced by antigens that are able to engage multiple FcεRIs on effector cells (Kanner and Metzger 1983; Scholl et al. 2005a; Segal et al. 1977), resulting in cross-linking of FcεRI and the release of anaphylactic mediators. Since IgE antibodies are tightly bound to their FcεRI receptor in a bent form (Gould et al. 2003), more then 2 antigen epitopes have to be displayed rather rigidly in a specific range of spatial conformation in order to cross-link IgE bound to FcεRI and induce its downstream effects (Holowka et al. 2007; Jensen-Jarolim et al. 2010). Molecules present in high density, such as certain overexpressed tumor antigens that are tightly packed on cell membranes or in lipid rafts, may cross-link FcεRI by forming tumor-associated molecular patterns, inducing a type I hypersensitivity reaction within the tumor microenvironment and resulting in tumor cytotoxicity (Jensen-Jarolim et al. 2010). Importantly, no systemic reactions have been observed in the various in vivo cancer models examined via a range of approaches by multiple independent groups, aimed to address IgE-mediated activity in cancer immunotherapy in the studies described above. Of particular concern are antigens shed or secreted by tumors such as ECD^{HER2}, FBP, and PSA. However, due to the mono-epitopic nature of the interaction with the IgE, these soluble antigens were not able to induce a type I hypersensitivity reaction.

A challenge in the field of IgE-based therapeutics is the lack of appropriate in vivo models to examine the efficacy and toxicity of the therapy [reviewed in Daniels et al. (2012b)]. Available models include immunocompetent murine models bearing tumor cell lines from the same genetic background (syngeneic), which can be used to evaluate the efficacy and potential toxicity of mouse IgE-mediated therapy and cancer vaccines, as well as the adaptive immune response induced by these murine antibodies. This model allows numerous administrations of therapeutic IgE without the potential neutralization of the therapy by a murine antibody immune response. However, due to the lack of species reactivity between human IgE and murine FcεRs, the anti-tumor effects of human IgE cannot be evaluated in syngeneic models. In addition, since FcεRI expression in the mouse is limited to MC and basophils, this model does not adequately reflect the immune response in humans where FcεRI is also expressed on other cells including APC and eosinophils. Furthermore, human tumors cannot be established due to their immunogenicity in mice. Xenograft models using immunocompromised mice allow establishment of human tumors and the use of human effector cells bearing both FcεRI and FcεRII/CD23 to evaluate the cytotoxic effects of a human antibody or a mouse IgE specific for a human antigen. However, these animals have impaired cellular and humoral immune responses and cannot be used to evaluate the adaptive immune response elicited by antibodies. In addition, human effector cells must be administered due to the lack of interaction of human IgE with mouse FcεRI.

Another disadvantage of this model is the limited supply of human effector cells and the possible lack of interaction of human cytokines secreted by these effector cells with the murine immune system. Establishment of a human FcεRIα transgenic mouse model in both BALB/c and C57BL/6 background (Dombrowicz et al. 1996, 1998; Fung-Leung et al. 1996), with murine FcεRIα knocked out and its human equivalent knocked in, allowed human IgE binding and expression of functional humanized FcεRI. This model also provides a constant repertoire of effector cells expressing human FcεRI mimicking that of humans for the study of human IgE and anaphylaxis (Dombrowicz et al. 1998; Fung-Leung et al. 1996; Kinet 1999) and the efficacy of molecules that inhibit anaphylaxis (Allen et al. 2007; Zhang et al. 2004; Zhu et al. 2002). In addition, it is useful for the evaluation of the anticancer effects mediated by FcεRI and the potential toxicity of IgE containing human constant regions. However, these animals are only transgenic for FcεRI and not FcεRII/CD23; therefore, the role of CD23 in the anticancer effects cannot be measured, which may underestimate the overall anticancer activity of the IgE. Additionally, human cancer cells cannot be established in this transgenic model, and the potential generation of an immune response against the foreign human components of the IgE antibodies may prevent multiple administrations of human antibodies.

The cynomolgus monkey (*M. fascicularis*) has been used for target validation and assessment of toxicity of therapeutic candidates designed to interfere with IgE biology, due to the highly conserved nature of FcεRIα between cynomolgus monkey and humans (Saul et al. 2014). This model has been used to study anaphylaxis induced by human IgE (Ishizaka et al. 1970; Weichman et al. 1982). Recent studies confirmed the cross-reactivity of human IgE with cynomolgus monkey effector cells and comparable binding kinetics to peripheral blood leukocytes from both species (Saul et al. 2014). However, human IgE was reported to dissociate faster from cynomolgus monkey PBMC compared with human, which was reflected in IgE-mediated ADCC assays, where higher concentrations of human IgE were needed to elicit an effector cell response of cynomolgus monkey cells (Saul et al. 2014). Moreover, IgE binding on PBMC yielded significantly different cytokine release profiles in each species, suggesting differences in downstream activation of effector cells. These data need to be considered when evaluating the toxicity of IgE-based therapies in this model.

Another animal model is the dog (*Canis lupus familiaris*), which has been used to examine IgE-mediated food allergy and atopic dermatitis (Ermel et al. 1997; Helm et al. 2003) and expresses FcεRI on MC (Brazis et al. 2002) and LC (Bonkobara et al. 2005). Spontaneous development of tumors also occurs in dogs, similar to humans (Singer et al. 2012). In addition, canine homologues of the tumor antigens EGFR and HER2/*neu* were susceptible to cetuximab and trastuzumab targeting, leading to growth arrest (Singer et al. 2012). These studies suggest that the canine may be a meaningful model to accurately assess the potential and safety of IgE-based immunotherapies against cancer. In summary, there are multiple animal models with distinct advantages and disadvantages that should be carefully considered when evaluating the efficacy and toxicity of IgE-based therapeutics.

Other evidences that support the safety of IgE as an anticancer therapy in humans include the presence of IgE antibodies against tumor antigens in patients with different types of cancer. Clinical and immunohistochemistry studies have decribed IgE that bind tumors both in the circulation and in tumor tissues (Fu et al. 2008; Neuchrist et al. 1994). Even so, anaphylaxis has not been observed in these studies on naturally occurring IgE antibodies in head and neck cancers, pancreatic, ovary, breast, or colon cancers (Jensen-Jarolim and Singer 2011). Importantly, clinical administration of exogenous IgE has been previously conducted without adverse events (Dreskin et al. 1987; Iio et al. 1978).

Allergic reactions are well-known clinical consequences of different types of treatments, including therapeutic proteins, chemotherapy agents, and antibodies, (Baldo 2013; Buttel et al. 2011; Romano et al. 2011; Singer and Jensen-Jarolim 2014). Pre-medication with antihistamines (Benadryl®) and/or corticosteroids (dexamethasone) are often administered prophylactically before chemotherapy (cisplatin, carboplatin) and monoclonal antibody treatment such as infliximab (Remicade®), trastuzumab (Herceptin®), and rituximab (Rituxan®) (Mezzano et al. 2014). In the unlikely event of signs of systemic anaphylaxis upon IgE treatment, epinephrine injection can be administered to neutralize this adverse effect (Simons et al. 2011). Established rapid drug desensitization procedures also exist to allow safe re-administration of a drug to patients that have developed allergic reactions to it and are performed when no alternative to the drug is deemed equally effective (Mezzano et al. 2014).

Despite the above statements, allergic reactions are relevant side effects that need to be carefully monitored. Neutralization and prevention strategies should be carefully considered. Safety measures and patient monitoring could include the evaluation of: (1) clinical signs of a systemic type I hypersensitivity (anaphylactic) reaction, (2) the presence of MC/basophils degranulation products in blood such as the elevation of histamine and tryptase, (3) the presence of autoantibodies to the targeted antigen and to the therapeutic antibody, and (4) blood concentrations of the targeted antigen. Correlation between blood antigen concentrations and clinical responsiveness, besides evaluation of toxicity in vivo, should also be examined. Treatment of patients with IgE-based therapeutics will have to proceed with extreme caution.

4 Concluding Thoughts

Several studies using IgE antibodies targeting various tumor antigens in different models have shown that these immunoglobulins have significant anticancer activity while being well-tolerated. The different properties of IgE compared to IgG support the development of IgE-based therapeutics to complement or enhance the antitumor activity shown by existing IgG-based therapeutics. Therefore, further studies including clinical trials are needed to confirm the efficacy and tolerability of IgE in cancer immunotherapy.

Acknowledgments Our work has been supported in part by grants from NIH/NCI: R41CA137881, R01CA136841, R01CA18115, K01CA138559, R21CA179680, and the Susan G. Komen Breast Cancer Foundation Basic, Clinical and Translational Research Grant BCTR0706771.

References

Achatz G, Achatz-Straussberger G, Feichtner S, Koenigsberger S, Lenz S, Peckl-Schmid D, Zaborsky N, Lamers M (2010) The biology of IgE: molecular mechanism restraining potentially dangerous high serum IgE titres in vivo In: Penichet ML, Jensen-Jarolim E (eds.) Cancer and IgE: introducing the concept of AllergoOncology. Springer, New York, pp 13–36

Adams CW, Allison DE, Flagella K, Presta L, Clarke J, Dybdal N, McKeever K, Sliwkowski MX (2006) Humanization of a recombinant monoclonal antibody to produce a therapeutic HER dimerization inhibitor, pertuzumab. Cancer Immunol Immunother 55:717–727

Ahn ER, Vogel CL (2011) Dual HER2-targeted approaches in HER2-positive breast cancer. Breast Cancer Res Treat 131:371–383

Allen LC, Kepley CL, Saxon A, Zhang K (2007) Modifications to an Fcgamma-Fcvarepsilon fusion protein alter its effectiveness in the inhibition of FcvarepsilonRI-mediated functions. J Allergy Clin Immunol 120:462–468

Baldo BA (2013) Adverse events to monoclonal antibodies used for cancer therapy: focus on hypersensitivity responses. Oncoimmunology 2:e26333

Banchereau J, Briere F, Caux C, Davoust J, Lebecque S, Liu YJ, Pulendran B, Palucka K (2000) Immunobiology of dendritic cells. Annu Rev Immunol 18:767–811

Beatty GL, Chiorean EG, Fishman MP, Saboury B, Teitelbaum UR, Sun W, Huhn RD, Song W, Li D, Sharp LL, Torigian DA, O'Dwyer PJ, Vonderheide RH (2011) CD40 agonists alter tumor stroma and show efficacy against pancreatic carcinoma in mice and humans. Science 331:1612–1616

Beavil AJ, Young RJ, Sutton BJ, Perkins SJ (1995) Bent domain structure of recombinant human IgE-Fc in solution by X-ray and neutron scattering in conjunction with an automated curve fitting procedure. Biochemistry 34:14449–14461

Behring E, Kitasato S (1890) Ueber das Zustandekommen der Diphtherie-Immunität und der Tetanus-Immunität bei Thieren. Dtsch Med Wochenschr 16:1113–1114

Berchuck A, Kamel A, Whitaker R, Kerns B, Olt G, Kinney R, Soper JT, Dodge R, Clarke-Pearson DL, Marks P et al (1990) Overexpression of HER-2/neu is associated with poor survival in advanced epithelial ovarian cancer. Cancer Res 50:4087–4091

Berlyn KA, Schultes B, Leveugle B, Noujaim AA, Alexander RB, Mann DL (2001) Generation of CD4(+) and CD8(+) T lymphocyte responses by dendritic cells armed with PSA/anti-PSA (antigen/antibody) complexes. Clin Immunol 101:276–283

Bettler B, Hofstetter H, Rao M, Yokoyama WM, Kilchherr F, Conrad DH (1989) Molecular structure and expression of the murine lymphocyte low-affinity receptor for IgE (Fc epsilon RII). Proc Natl Acad Sci USA 86:7566–7570

Bieber T, de la Salle H, Wollenberg A, Hakimi J, Chizzonite R, Ring J, Hanau D, de la Salle C (1992) Human epidermal Langerhans cells express the high affinity receptor for immunoglobulin E (Fc epsilon RI). J Exp Med 175:1285–1290

Bonkobara M, Miyake F, Yagihara H, Yamada O, Azakami D, Washizu T, Cruz PD Jr, Ariizumi K (2005) Canine epidermal langerhans cells express alpha and gamma but not beta chains of high-affinity IgE receptor. Vet Res Commun 29:499–505

Boross P, Lohse S, Nederend M, Jansen JH, van Tetering G, Dechant M, Peipp M, Royle L, Liew LP, Boon L, van Rooijen N, Bleeker WK, Parren PW, van de Winkel JG, Valerius T, Leusen JH (2013) IgA EGFR antibodies mediate tumour killing in vivo. EMBO Mol Med 5:1213–1226

Braly P, Nicodemus CF, Chu C, Collins Y, Edwards R, Gordon A, McGuire W, Schoonmaker C, Whiteside T, Smith LM, Method M (2009) The immune adjuvant properties of front-line carboplatin-paclitaxel: a randomized phase 2 study of alternative schedules of intravenous oregovomab chemoimmunotherapy in advanced ovarian cancer. J Immunother 32:54–65

Brazis P, De Mora F, Ferrer L, Puigdemont A (2002) IgE enhances Fc epsilon RI expression and IgE-dependent TNF-alpha release from canine skin mast cells. Vet Immunol Immunopathol 85:205–212

Buttel IC, Chamberlain P, Chowers Y, Ehmann F, Greinacher A, Jefferis R, Kramer D, Kropshofer H, Lloyd P, Lubiniecki A, Krause R, Mire-Sluis A, Platts-Mills T, Ragheb JA, Reipert BM, Schellekens H, Seitz R, Stas P, Subramanyam M, Thorpe R, Trouvin JH, Weise M, Windisch J, Schneider CK (2011) Taking immunogenicity assessment of therapeutic proteins to the next level. Biologicals 39:100–109

Cameron F, McCormack PL (2014) Obinutuzumab: first global approval. Drugs 74:147–154

Campoli M, Ferris R, Ferrone S, Wang X (2010) Immunotherapy of malignant disease with tumor antigen-specific monoclonal antibodies. Clin Cancer Res 16:11–20

Capron M, Capron A (1994) Immunoglobulin E and effector cells in schistosomiasis. Science 264:1876–1877

Capron A, Dessaint JP (1985) Effector and regulatory mechanisms in immunity to schistosomes: a heuristic view. Annu Rev Immunol 3:455–476

Capron A, Dombrowicz D, Capron M (1999) Regulation of the immune response in experimental and human schistosomiasis: the limits of an attractive paradigm. Microbes Infect 1:485–490

Caruso RA, Parisi A, Quattrocchi E, Scardigno M, Branca G, Parisi C, Luciano R, Paparo D, Fedele F (2011) Ultrastructural descriptions of heterotypic aggregation between eosinophils and tumor cells in human gastric carcinomas. Ultrastruct Pathol 35:145–149

Chu RS, Targoni OS, Krieg AM, Lehmann PV, Harding CV (1997) CpG oligodeoxynucleotides act as adjuvants that switch on T helper 1 (Th1) immunity. J Exp Med 186:1623–1631

Clynes RA, Towers TL, Presta LG, Ravetch JV (2000) Inhibitory Fc receptors modulate in vivo cytotoxicity against tumor targets. Nat Med 6:443–446

Conrad DH (1990) Fc epsilon RII/CD23: the low affinity receptor for IgE. Annu Rev Immunol 8:623–645

Cook J, Hagemann T (2013) Tumour-associated macrophages and cancer. Curr Opin Pharmacol 13:595–601

Cooper PJ, Ayre G, Martin C, Rizzo JA, Ponte EV, Cruz AA (2008) Geohelminth infections: a review of the role of IgE and assessment of potential risks of anti-IgE treatment. Allergy 63:409–417

Cormier SA, Taranova AG, Bedient C, Nguyen T, Protheroe C, Pero R, Dimina D, Ochkur SI, O'Neill K, Colbert D, Lombari TR, Constant S, McGarry MP, Lee JJ, Lee NA (2006) Pivotal advance: eosinophil infiltration of solid tumors is an early and persistent inflammatory host response. J Leukoc Biol 79:1131–1139

Cuschieri A, Talbot IC, Weeden S (2002) Influence of pathological tumour variables on long-term survival in resectable gastric cancer. Br J Cancer 86:674–679

Dalton DK, Noelle RJ (2012) The roles of mast cells in anticancer immunity. Cancer Immunol Immunother 61:1511–1520

Daniels TR, Rodriguez JA, Ortiz-Sanchez E, Helguera G, Penichet ML (2010) The IgE antibody and its use in cancer immunotherapy In: Penichet ML, Jensen-Jarolim E (eds.) Cancer and IgE: introducing the concept of AllergoOncology. Springer, New York, pp 159–183

Daniels TR, Leuchter RK, Quintero R, Helguera G, Rodriguez JA, Martinez-Maza O, Schultes BC, Nicodemus CF, Penichet ML (2012a) Targeting HER2/neu with a fully human IgE to harness the allergic reaction against cancer cells. Cancer Immunol Immunother 61:991–1003

Daniels TR, Martinez-Maza O, Penichet ML (2012b) Animal models for IgE-meditated cancer immunotherapy. Cancer Immunol Immunother 61:1535–1546

Daniels-Wells TR, Helguera G, Leuchter RK, Quintero R, Kozman M, Rodriguez JA, Ortiz-Sanchez E, Martinez-Maza O, Schultes BC, Nicodemus CF, Penichet ML (2013) A novel IgE antibody targeting the prostate-specific antigen as a potential prostate cancer therapy. BMC Cancer 13:195–207

de Andres B, Rakasz E, Hagen M, McCormik ML, Mueller AL, Elliot D, Metwali A, Sandor M, Britigan BE, Weinstock JV, Lynch RG (1997) Lack of Fc-epsilon receptors on murine eosinophils: implications for the functional significance of elevated IgE and eosinophils in parasitic infections. Blood 89:3826–3836

de Vries VC, Wasiuk A, Bennett KA, Benson MJ, Elgueta R, Waldschmidt TJ, Noelle RJ (2009) Mast cell degranulation breaks peripheral tolerance. Am J Transplant 9:2270–2280

Delespesse G, Suter U, Mossalayi D, Bettler B, Sarfati M, Hofstetter H, Kilcherr E, Debre P, Dalloul A (1991) Expression, structure, and function of the CD23 antigen. Adv Immunol 49:149–191

della Rovere F, Granata A, Familiari D, D'Arrigo G, Mondello B, Basile G (2007) Mast cells in invasive ductal breast cancer: different behavior in high and minimum hormone-receptive cancers. Anticancer Res 27:2465–2471

Dombrowicz D, Brini AT, Flamand V, Hicks E, Snouwaert JN, Kinet JP, Koller BH (1996) Anaphylaxis mediated through a humanized high affinity IgE receptor. J Immunol 157:1645–1651

Dombrowicz D, Lin S, Flamand V, Brini AT, Koller BH, Kinet JP (1998) Allergy-associated FcRbeta is a molecular amplifier of IgE- and IgG-mediated in vivo responses. Immunity 8:517–529

Dombrowicz D, Quatannens B, Papin JP, Capron A, Capron M (2000) Expression of a functional Fc epsilon RI on rat eosinophils and macrophages. J Immunol 165:1266–1271

Dorta RG, Landman G, Kowalski LP, Lauris JR, Latorre MR, Oliveira DT (2002) Tumour-associated tissue eosinophilia as a prognostic factor in oral squamous cell carcinomas. Histopathology 41:152–157

Dreskin SC, Goldsmith PK, Strober W, Zech LA, Gallin JI (1987) Metabolism of immunoglobulin E in patients with markedly elevated serum immunoglobulin E levels. J Clin Invest 79:1764–1772

Dunne DW, Butterworth AE, Fulford AJ, Kariuki HC, Langley JG, Ouma JH, Capron A, Pierce RJ, Sturrock RF (1992) Immunity after treatment of human schistosomiasis: association between IgE antibodies to adult worm antigens and resistance to reinfection. Eur J Immunol 22:1483–1494

Ehrlich P (1891) Experimentelle Untersuchungen über Immunität. Dtsch Med Wochenschr 17:976

Ehrlich P (1901a) Die schutzstoffe des blutes. Dtsch Med Wochenschr 27:865

Ehrlich P (1901b) Die seitenkettentheorie und ihre gegner. Münch Med Wochenschr 18:2123

Ermel RW, Kock M, Griffey SM, Reinhart GA, Frick OL (1997) The atopic dog: a model for food allergy. Lab Anim Sci 47:40–49

Evans-Axelsson S, Ulmert D, Orbom A, Peterson P, Nilsson O, Wennerberg J, Strand J, Wingardh K, Olsson T, Hagman Z, Tolmachev V, Bjartell A, Lilja H, Strand SE (2012) Targeting free prostate-specific antigen for in vivo imaging of prostate cancer using a monoclonal antibody specific for unique epitopes accessible on free prostate-specific antigen alone. Cancer Biother Radiopharm 27:243–251

Fernandez Y, Cueva J, Palomo AG, Ramos M, de Juan A, Calvo L, Garcia-Mata J, Garcia-Teijido P, Pelaez I, Garcia-Estevez L (2010) Novel therapeutic approaches to the treatment of metastatic breast cancer. Cancer Treat Rev 36:33–42

Fernandez-Acenero MJ, Galindo-Gallego M, Sanz J, Aljama A (2000) Prognostic influence of tumor-associated eosinophilic infiltrate in colorectal carcinoma. Cancer 88:1544–1548

Finkelman FD, Holmes J, Katona IM, Urban JF Jr, Beckmann MP, Park LS, Schooley KA, Coffman RL, Mosmann TR, Paul WE (1990) Lymphokine control of in vivo immunoglobulin isotype selection. Annu Rev Immunol 8:303–333

Fu SL, Pierre J, Smith-Norowitz TA, Hagler M, Bowne W, Pincus MR, Mueller CM, Zenilman ME, Bluth MH (2008) Immunoglobulin E antibodies from pancreatic cancer patients mediate antibody-dependent cell-mediated cytotoxicity against pancreatic cancer cells. Clin Exp Immunol 153:401–409

Fung-Leung WP, De Sousa-Hitzler J, Ishaque A, Zhou L, Pang J, Ngo K, Panakos JA, Chourmouzis E, Liu FT, Lau CY (1996) Transgenic mice expressing the human high-affinity immunoglobulin (Ig) E receptor alpha chain respond to human IgE in mast cell degranulation and in allergic reactions. J Exp Med 183:49–56

Galli SJ, Tsai M (2010) Mast cells in allergy and infection: versatile effector and regulatory cells in innate and adaptive immunity. Eur J Immunol 40:1843–1851

Galli SJ, Grimbaldeston M, Tsai M (2008) Immunomodulatory mast cells: negative, as well as positive, regulators of immunity. Nat Rev Immunol 8:478–486

Gatault S, Legrand F, Delbeke M, Loiseau S, Capron M (2012) Involvement of eosinophils in the anti-tumor response. Cancer Immunol Immunother 61:1527–1534

Gould HJ, Sutton BJ (2008) IgE in allergy and asthma today. Nat Rev Immunol 8:205–217

Gould HJ, Mackay GA, Karagiannis SN, O'Toole CM, Marsh PJ, Daniel BE, Coney LR, Zurawski VR Jr, Joseph M, Capron M, Gilbert M, Murphy GF, Korngold R (1999) Comparison of IgE and IgG antibody-dependent cytotoxicity in vitro and in a SCID mouse xenograft model of ovarian carcinoma. Eur J Immunol 29:3527–3537

Gould HJ, Sutton BJ, Beavil AJ, Beavil RL, McCloskey N, Coker HA, Fear D, Smurthwaite L (2003) The biology of IgE and the basis of allergic disease. Annu Rev Immunol 21:579–628

Gourevitch MM, von Mensdorff-Pouilly S, Litvinov SV, Kenemans P, van Kamp GJ, Verstraeten AA, Hilgers J (1995) Polymorphic epithelial mucin (MUC-1)-containing circulating immune complexes in carcinoma patients. Br J Cancer 72:934–938

Grillo-Lopez AJ (2000) Rituximab: an insider's historical perspective. Semin Oncol 27:9–16

Guttinger M, Guidi F, Chinol M, Reali E, Veglia F, Viale G, Paganelli G, Corti A, Siccardi AG (2000) Adoptive immunotherapy by avidin-driven cytotoxic T lymphocyte-tumor bridging. Cancer Res 60:4211–4215

Hakimi J, Seals C, Kondas JA, Pettine L, Danho W, Kochan J (1990) The alpha subunit of the human IgE receptor (FcERI) is sufficient for high affinity IgE binding. J Biol Chem 265:22079–22081

Helm RM, Ermel RW, Frick OL (2003) Nonmurine animal models of food allergy. Environ Health Perspect 111:239–244

Hibbert RG, Teriete P, Grundy GJ, Beavil RL, Reljic R, Holers VM, Hannan JP, Sutton BJ, Gould HJ, McDonnell JM (2005) The structure of human CD23 and its interactions with IgE and CD21. J Exp Med 202:751–760

Holowka D, Sil D, Torigoe C, Baird B (2007) Insights into immunoglobulin E receptor signaling from structurally defined ligands. Immunol Rev 217:269–279

Hudis CA (2007) Trastuzumab–mechanism of action and use in clinical practice. N Engl J Med 357:39–51

Iio A, Waldmann TA, Strober W (1978) Metabolic study of human IgE: evidence for an extravascular catabolic pathway. J Immunol 120:1696–1701

Ishibashi S, Ohashi Y, Suzuki T, Miyazaki S, Moriya T, Satomi S, Sasano H (2006) Tumor-associated tissue eosinophilia in human esophageal squamous cell carcinoma. Anticancer Res 26:1419–1424

Ishizaka T, Ishizaka K, Orange RP, Austen KF (1970) The capacity of human immunoglobulin E to mediate the release of histamine and slow reacting substance of anaphylaxis (SRS-A) from monkey lung. J Immunol 104:335–343

Janeway CA, Travers P, Walport M, Schlomchick M (2005a) Chapter 4: the generation of lymphocyte antigen receptors. Immunobiology: the immune system in health and disease. Garland Science, New York, pp 135–168

Janeway CA, Travers P, Walport M, Schlomchick M (2005b) Chapter 8: T Cell-mediated immunity. Immunobiology: the immune system in health and disease. Garland Science, New York, pp 319–366

Janeway CA, Travers P, Walport M, Schlomchick M (2005c) Chapter 12: allergy and hypersensitivity. Immunobiology: the immune system in health and disease. Garland Science, New York, pp 517–555

Janeway CA, Travers P, Walport M, Shlomchik M (2005d) Chapter 9: The humoral immune response. Immunobiology: the immune system in health and disease. Garland Science, New York, pp 367–406

Jensen-Jarolim E, Singer J (2011) Why could passive immunoglobulin E antibody therapy be safe in clinical oncology? Clin Exp Allergy 41:1337–1340

Jensen-Jarolim E, Mechtcheriakova D, Pali-Schoell I (2010) The targets of IgE: allergen-associated and tumor-associated molecular patterns. In: Penichet ML, Jensen-Jarolim E (eds.) Cancer and IgE: introducing the concept of AllergoOncology. Springer, New York, pp 47–78

Johansson A, Rudolfsson S, Hammarsten P, Halin S, Pietras K, Jones J, Stattin P, Egevad L, Granfors T, Wikstrom P, Bergh A (2010) Mast cells are novel independent prognostic markers in prostate cancer and represent a target for therapy. Am J Pathol 177:1031–1041

Josephs DH, Spicer JF, Corrigan CJ, Gould HJ, Karagiannis SN (2013) Epidemiological associations of allergy, IgE and cancer. Clin Exp Allergy 43:1110–1123

Junutula JR, Flagella KM, Graham RA, Parsons KL, Ha E, Raab H, Bhakta S, Nguyen T, Dugger DL, Li G, Mai E, Lewis Phillips GD, Hiraragi H, Fuji RN, Tibbitts J, Vandlen R, Spencer SD, Scheller RH, Polakis P, Sliwkowski MX (2010) Engineered thio-trastuzumab-DM1 conjugate with an improved therapeutic index to target human epidermal growth factor receptor 2-positive breast cancer. Clin Cancer Res 16:4769–4778

Kanner BI, Metzger H (1983) Crosslinking of the receptors for immunoglobulin E depolarizes the plasma membrane of rat basophilic leukemia cells. Proc Natl Acad Sci USA 80:5744–5748

Karagiannis SN, Wang Q, East N, Burke F, Riffard S, Bracher MG, Thompson RG, Durham SR, Schwartz LB, Balkwill FR, Gould HJ (2003) Activity of human monocytes in IgE antibody-dependent surveillance and killing of ovarian tumor cells. Eur J Immunol 33:1030–1040

Karagiannis SN, Bracher MG, Hunt J, McCloskey N, Beavil RL, Beavil AJ, Fear DJ, Thompson RG, East N, Burke F, Moore RJ, Dombrowicz DD, Balkwill FR, Gould HJ (2007) IgE-antibody-dependent immunotherapy of solid tumors: cytotoxic and phagocytic mechanisms of eradication of ovarian cancer cells. J Immunol 179:2832–2843

Karagiannis SN, Bracher MG, Beavil RL, Beavil AJ, Hunt J, McCloskey N, Thompson RG, East N, Burke F, Sutton BJ, Dombrowicz D, Balkwill FR, Gould HJ (2008) Role of IgE receptors in IgE antibody-dependent cytotoxicity and phagocytosis of ovarian tumor cells by human monocytic cells. Cancer Immunol Immunother 57:247–263

Karagiannis P, Singer J, Hunt J, Gan SK, Rudman SM, Mechtcheriakova D, Knittelfelder R, Daniels TR, Hobson PS, Beavil AJ, Spicer J, Nestle FO, Penichet ML, Gould HJ, Jensen-Jarolim E, Karagiannis SN (2009) Characterisation of an engineered trastuzumab IgE antibody and effector cell mechanisms targeting HER2/neu-positive tumour cells. Cancer Immunol Immunother 58:915–930

Karagiannis SN, Josephs DH, Karagiannis P, Gilbert AE, Saul L, Rudman SM, Dodev T, Koers A, Blower PJ, Corrigan C, Beavil AJ, Spicer JF, Nestle FO, Gould HJ (2012) Recombinant IgE antibodies for passive immunotherapy of solid tumours: from concept towards clinical application. Cancer Immunol Immunother 61:1547–1564

Katzenwadel A, Schleer H, Gierschner D, Wetterauer U, Elsasser-Beile U (2000) Construction and in vivo evaluation of an anti-PSA x anti-CD3 bispecific antibody for the immunotherapy of prostate cancer. Anticancer Res 20:1551–1555

Kellner C, Derer S, Valerius T, Peipp M (2014) Boosting ADCC and CDC activity by Fc engineering and evaluation of antibody effector functions. Methods 65:105–113

Kershaw MH, Darcy PK, Trapani JA, Smyth MJ (1996) The use of chimeric human Fc(epsilon) receptor I to redirect cytotoxic T lymphocytes to tumors. J Leukoc Biol 60:721–728

Kershaw MH, Darcy PK, Trapani JA, MacGregor D, Smyth MJ (1998) Tumor-specific IgE-mediated inhibition of human colorectal carcinoma xenograft growth. Oncol Res 10:133–142

Kilmon MA, Ghirlando R, Strub MP, Beavil RL, Gould HJ, Conrad DH (2001) Regulation of IgE production requires oligomerization of CD23. J Immunol 167:3139–3145

Kinet JP (1999) The high-affinity IgE receptor (Fc epsilon RI): from physiology to pathology. Annu Rev Immunol 17:931–972

King DM, Albertini MR, Schalch H, Hank JA, Gan J, Surfus J, Mahvi D, Schiller JH, Warner T, Kim K, Eickhoff J, Kendra K, Reisfeld R, Gillies SD, Sondel P (2004) Phase I clinical trial of the immunocytokine EMD 273063 in melanoma patients. J Clin Oncol 22:4463–4473

Kohler G, Milstein C (1975) Continuous cultures of fused cells secreting antibody of predefined specificity. Nature 256:495–497

Kotera Y, Fontenot JD, Pecher G, Metzgar RS, Finn OJ (1994) Humoral immunity against a tandem repeat epitope of human mucin MUC-1 in sera from breast, pancreatic, and colon cancer patients. Cancer Res 54:2856–2860

Lee HW, Choi HJ, Ha SJ, Lee KT, Kwon YG (2013) Recruitment of monocytes/macrophages in different tumor microenvironments. Biochim Biophys Acta 1835:170–179

Leget GA, Czuczman MS (1998) Use of rituximab, the new FDA-approved antibody. Curr Opin Oncol 10:548–551

Legrand F, Driss V, Delbeke M, Loiseau S, Hermann E, Dombrowicz D, Capron M (2010) Human eosinophils exert TNF-alpha and granzyme A-mediated tumoricidal activity toward colon carcinoma cells. J Immunol 185:7443–7451

Lemos MP, Fan L, Lo D, Laufer TM (2003) CD8alpha+ and CD11b+ dendritic cell-restricted MHC class II controls Th1 CD4+T cell immunity. J Immunol 171:5077–5084

Lewin J, Thomas D (2013) Denosumab: a new treatment option for giant cell tumor of bone. Drugs Today (Barc) 49:693–700

Liu AY, Robinson RR, Murray ED Jr, Ledbetter JA, Hellstrom I, Hellstrom KE (1987) Production of a mouse-human chimeric monoclonal antibody to CD20 with potent Fc-dependent biologic activity. J Immunol 139:3521–3526

MacGlashan D Jr, Lichtenstein LM, McKenzie-White J, Chichester K, Henry AJ, Sutton BJ, Gould HJ (1999) Upregulation of FcepsilonRI on human basophils by IgE antibody is mediated by interaction of IgE with FcepsilonRI. J Allergy Clin Immunol 104:492–498

Manz RA, Hauser AE, Hiepe F, Radbruch A (2005) Maintenance of serum antibody levels. Annu Rev Immunol 23:367–386

Martinelli E, De Palma R, Orditura M, De Vita F, Ciardiello F (2009) Anti-epidermal growth factor receptor monoclonal antibodies in cancer therapy. Clin Exp Immunol 158:1–9

Martinez FO, Gordon S (2014) The M1 and M2 paradigm of macrophage activation: time for reassessment. F1000Prime Rep 6, 13

Martínez-Maza O, Moreno AD, Cozen W (2010) Epidemiological evidence: IgE, allergies, and hematopoietic malignancies. In: Penichet ML, Jensen-Jarolim E (eds.) Cancer and IgE: introducing the concept of AllergoOncology. Springer, New York, pp 79–136

Mathur A, Conrad DH, Lynch RG (1988) Characterization of the murine T cell receptor for IgE (Fc epsilon RII). Demonstration of shared and unshared epitopes with the B cell Fc epsilon RII. J Immunol 141:2661–2667

Matta GM, Battaglio S, Dibello C, Napoli P, Baldi C, Ciccone G, Coscia M, Boccadoro M, Massaia M (2007) Polyclonal immunoglobulin E levels are correlated with hemoglobin values and overall survival in patients with multiple myeloma. Clin Cancer Res 13:5348–5354

Mattes J, Hulett M, Xie W, Hogan S, Rothenberg ME, Foster P, Parish C (2003) Immunotherapy of cytotoxic T cell-resistant tumors by T helper 2 cells: an eotaxin and STAT6-dependent process. J Exp Med 197:387–393

Maurer D, Fiebiger S, Ebner C, Reininger B, Fischer GF, Wichlas S, Jouvin MH, Schmitt-Egenolf M, Kraft D, Kinet JP, Stingl G (1996) Peripheral blood dendritic cells express Fc epsilon RI as a complex composed of Fc epsilon RI alpha- and Fc epsilon RI gamma-chains and can use this receptor for IgE-mediated allergen presentation. J Immunol 157:607–616

McCloskey N, Hunt J, Beavil RL, Jutton MR, Grundy GJ, Girardi E, Fabiane SM, Fear DJ, Conrad DH, Sutton BJ, Gould HJ (2007) Soluble CD23 monomers inhibit and oligomers stimulate IgE synthesis in human B cells. J Biol Chem 282:24083–24091

McDonnell JM, Calvert R, Beavil RL, Beavil AJ, Henry AJ, Sutton BJ, Gould HJ, Cowburn D (2001) The structure of the IgE Cepsilon2 domain and its role in stabilizing the complex with its high-affinity receptor FcepsilonRIalpha. Nat Struct Biol 8:437–441

Meden H, Marx D, Rath W, Kron M, Fattahi-Meibodi A, Hinney B, Kuhn W, Schauer A (1994) Overexpression of the oncogene c-erb B2 in primary ovarian cancer: evaluation of the prognostic value in a Cox proportional hazards multiple regression. Int J Gynecol Pathol 13:45–53

Mezzano V, Giavina-Bianchi P, Picard M, Caiado J, Castells M (2014) Drug desensitization in the management of hypersensitivity reactions to monoclonal antibodies and chemotherapy. BioDrugs 28:133–144

Mount PF, Sutton VR, Li W, Burgess J, Mc KIF, Pietersz GA, Trapani JA (1994) Chimeric (mouse/human) anti-colon cancer antibody c30.6 inhibits the growth of human colorectal cancer xenografts in scid/scid mice. Cancer Res 54:6160–6166

Murphy KP (2012) Chapter 5: the generation of lymphocyte antigen receptors. Janeway's immunobiology. Garland Science, New York, pp 157–200

Munitz A, Levi-Schaffer F (2004) Eosinophils: 'new' roles for 'old' cells. Allergy 59:268–275

Nagy E, Berczi I, Sehon AH (1991) Growth inhibition of murine mammary carcinoma by monoclonal IgE antibodies specific for the mammary tumor virus. Cancer Immunol Immunother 34:63–69

Nahta R, Shabaya S, Ozbay T, Rowe DL (2009) Personalizing HER2-targeted therapy in metastatic breast cancer beyond HER2 status: what we have learned from clinical specimens. Curr Pharmacogenomics Person Med 7:263–274

Neuchrist C, Kornfehl J, Grasl M, Lassmann H, Kraft D, Ehrenberger K, Scheiner O (1994) Distribution of immunoglobulins in squamous cell carcinoma of the head and neck. Int Arch Allergy Immunol 104:97–100

Nigro EA, Brini AT, Soprana E, Ambrosi A, Dombrowicz D, Siccardi AG, Vangelista L (2009) Antitumor IgE adjuvanticity: key role of Fc epsilon RI. J Immunol 183:4530–4536

Nigro EA, Soprana E, Brini AT, Ambrosi A, Yenagi VA, Dombrowicz D, Siccardi AG, Vangelista L (2012) An antitumor cellular vaccine based on a mini-membrane IgE. J Immunol 188:103–110

Nimmerjahn F, Ravetch JV (2007) Antibodies, Fc receptors and cancer. Curr Opin Immunol 19:239–245

Novak N, Kraft S, Haberstok J, Geiger E, Allam P, Bieber T (2002) A reducing microenvironment leads to the generation of Fc epsilon RI high inflammatory dendritic epidermal cells (IDEC). J Invest Dermatol 119:842–849

Novak N, Gros E, Bieber T, Allam JP (2010) Human skin and oral mucosal dendritic cells as 'good guys' and 'bad guys' in allergic immune responses. Clin Exp Immunol 161:28–33

Obata-Ninomiya K, Ishiwata K, Tsutsui H, Nei Y, Yoshikawa S, Kawano Y, Minegishi Y, Ohta N, Watanabe N, Kanuka H, Karasuyama H (2013) The skin is an important bulwark of acquired immunity against intestinal helminths. Exp Med 210:2583–2595

Obort AS, Ajadi MB, Akinloye O (2013) Prostate-specific antigen: any successor in sight? Rev Urol 15:97–107

Osterhoff B, Rappersberger K, Wang B, Koszik F, Ochiai K, Kinet JP, Stingl G (1994) Immunomorphologic characterization of Fc epsilon RI-bearing cells within the human dermis. J Invest Dermatol 102:315–320

Ovchinnikov DA (2008) Macrophages in the embryo and beyond: much more than just giant phagocytes. Genesis 46:447–462

Pegram M, Ngo D (2006) Application and potential limitations of animal models utilized in the development of trastuzumab (Herceptin): a case study. Adv Drug Deliv Rev 58:723–734

Peipp M, Dechant M, Valerius T (2008) Effector mechanisms of therapeutic antibodies against ErbB receptors. Curr Opin Immunol 20:436–443

Penichet ML, Jensen-Jarolim E (2010) (eds.) Cancer and IgE: introducing the concept of AllergoOncology. Springer, New York, USA

Preithner S, Elm S, Lippold S, Locher M, Wolf A, da Silva AJ, Baeuerle PA, Prang NS (2006) High concentrations of therapeutic IgG1 antibodies are needed to compensate for inhibition of antibody-dependent cellular cytotoxicity by excess endogenous immunoglobulin G. Mol Immunol 43:1183–1193

Ravetch JV, Kinet JP (1991) Fc receptors. Annu Rev Immunol 9:457–492

Reali E, Greiner JW, Corti A, Gould HJ, Bottazzoli F, Paganelli G, Schlom J, Siccardi AG (2001) IgEs targeted on tumor cells: therapeutic activity and potential in the design of tumor vaccines. Cancer Res 61:5517–5522

Reff ME, Carner K, Chambers KS, Chinn PC, Leonard JE, Raab R, Newman RA, Hanna N, Anderson DR (1994) Depletion of B cells in vivo by a chimeric mouse human monoclonal antibody to CD20. Blood 83:435–445

Rezvani AR, Maloney DG (2011) Rituximab resistance. Best Pract Res Clin Haematol 24:203–216

Rieger A, Wang B, Kilgus O, Ochiai K, Mauerer D, Fodinger D, Kinet JP, Stingl G (1992) Fc epsilon RI mediates IgE binding to human epidermal Langerhans cells. J Invest Dermatol 99:30S–32S

Riemer AB, Untersmayr E, Knittelfelder R, Duschl A, Pehamberger H, Zielinski CC, Scheiner O, Jensen-Jarolim E (2007) Active induction of tumor-specific IgE antibodies by oral mimotope vaccination. Cancer Res 67:3406–3411

Romano A, Torres MJ, Castells M, Sanz ML, Blanca M (2011) Diagnosis and management of drug hypersensitivity reactions. J Allergy Clin Immunol 127:S67–S73

Roulois D, Gregoire M, Fonteneau JF (2013) MUC1-specific cytotoxic T lymphocytes in cancer therapy: induction and challenge. Biomed Res Int 2013:871936–871946

Rudman SM, Josephs DH, Cambrook H, Karagiannis P, Gilbert AE, Dodev T, Hunt J, Koers A, Montes A, Taams L, Canevari S, Figini M, Blower PJ, Beavil AJ, Nicodemus CF, Corrigan C, Kaye SB, Nestle FO, Gould HJ, Spicer JF, Karagiannis SN (2011) Harnessing engineered antibodies of the IgE class to combat malignancy: initial assessment of FcvarepsilonRI-mediated basophil activation by a tumour-specific IgE antibody to evaluate the risk of type I hypersensitivity. Clin Exp Allergy 41:1400–1413

Salomon DS, Brandt R, Ciardiello F, Normanno N (1995) Epidermal growth factor-related peptides and their receptors in human malignancies. Crit Rev Oncol Hematol 19:183–232

Sanderson CJ (1988) Interleukin-5: an eosinophil growth and activation factor. Dev Biol Stand 69:23–29

Saul L, Josephs DH, Cutler K, Bradwell A, Karagiannis P, Selkirk C, Gould HJ, Jones P, Spicer JF, Karagiannis SN (2014) Comparative reactivity of human IgE to cynomolgus monkey and human effector cells and effects on IgE effector cell potency. MAbs 6:509–522

Schier R, Marks JD, Wolf EJ, Apell G, Wong C, McCartney JE, Bookman MA, Huston JS, Houston LL, Weiner LM, Adams GP (1995) In vitro and in vivo characterization of a human anti-c-erbB-2 single-chain Fv isolated from a filamentous phage antibody library. Immunotechnology 1:73–81

Schier R, McCall A, Adams GP, Marshall KW, Merritt H, Yim M, Crawford RS, Weiner LM, Marks C, Marks JD (1996) Isolation of picomolar affinity anti-c-erbB-2 single-chain Fv by molecular evolution of the complementarity determining regions in the center of the antibody binding site. J Mol Biol 263:551–567

Schmiedel J, Blaukat A, Li S, Knochel T, Ferguson KM (2008) Matuzumab binding to EGFR prevents the conformational rearrangement required for dimerization. Cancer Cell 13:365–373

Scholl I, Kalkura N, Shedziankova Y, Bergmann A, Verdino P, Knittelfelder R, Kopp T, Hantusch B, Betzel C, Dierks K, Scheiner O, Boltz-Nitulescu G, Keller W, Jensen-Jarolim E (2005a) Dimerization of the major birch pollen allergen Bet v 1 is important for its in vivo IgE-cross-linking potential in mice. J Immunol 175:6645–6650

Scholl I, Untersmayr E, Bakos N, Roth-Walter F, Gleiss A, Boltz-Nitulescu G, Scheiner O, Jensen-Jarolim E (2005b) Antiulcer drugs promote oral sensitization and hypersensitivity to hazelnut allergens in BALB/c mice and humans. Am J Clin Nutr 81:154–160

Segal DM, Taurog JD, Metzger H (1977) Dimeric immunoglobulin E serves as a unit signal for mast cell degranulation. Proc Natl Acad Sci USA 74:2993–2997

Simons FE, Ardusso LR, Bilo MB, El-Gamal YM, Ledford DK, Ring J, Sanchez-Borges M, Senna GE, Sheikh A, Thong BY (2011) World allergy organization anaphylaxis guidelines: summary. J Allergy Clin Immunol 127:587–593

Simson L, Ellyard JI, Dent LA, Matthaei KI, Rothenberg ME, Foster PS, Smyth MJ, Parish CR (2007) Regulation of carcinogenesis by IL-5 and CCL11: a potential role for eosinophils in tumor immune surveillance. J Immunol 178:4222–4229

Singer J, Jensen-Jarolim E (2014) IgE-based immunotherapy of cancer: challenges and chances. Allergy 69:137–149

Singer J, Weichselbaumer M, Stockner T, Mechtcheriakova D, Sobanov Y, Bajna E, Wrba F, Horvat R, Thalhammer JG, Willmann M, Jensen-Jarolim E (2012) Comparative oncology: ErbB-1 and ErbB-2 homologues in canine cancer are susceptible to cetuximab and trastuzumab targeting. Mol Immunol 50:200–209

Sinha AA, Sackrison JL, DeLeon OF, Wilson MJ, Gleason DF (1996) Antibody immunoglobulin G (IgG) against human prostatic specific antigen (PSA) as a carrier protein for chemotherapeutic drugs to human prostate tumors: part 1 a double immunofluorescence analysis. Anat Rec 245:652–661

Sinha AA, Quast BJ, Reddy PK, Elson MK, Wilson MJ (1999) Intravenous injection of an immunoconjugate (anti-PSA-IgG conjugated to 5-fluoro-2'-deoxyuridine) selectively inhibits cell proliferation and induces cell death in human prostate cancer cell tumors grown in nude mice. Anticancer Res 19:893–902

Slamon DJ, Clark GM, Wong SG, Levin WJ, Ullrich A, McGuire WL (1987) Human breast cancer: correlation of relapse and survival with amplification of the HER-2/neu oncogene. Science 235:177–182

Slamon DJ, Godolphin W, Jones LA, Holt JA, Wong SG, Keith DE, Levin WJ, Stuart SG, Udove J, Ullrich A et al (1989) Studies of the HER-2/neu proto-oncogene in human breast and ovarian cancer. Science 244:707–712

Slamon DJ, Leyland-Jones B, Shak S, Fuchs H, Paton V, Bajamonde A, Fleming T, Eiermann W, Wolter J, Pegram M, Baselga J, Norton L (2001) Use of chemotherapy plus a monoclonal antibody against HER2 for metastatic breast cancer that overexpresses HER2. N Engl J Med 344:783–792

Sliwkowski MX, Mellman I (2013) Antibody therapeutics in cancer. Science 341:1192–1198

Smith MR (2003) Rituximab (monoclonal anti-CD20 antibody): mechanisms of action and resistance. Oncogene 22:7359–7368

Soucek L, Lawlor ER, Soto D, Shchors K, Swigart LB, Evan GI (2007) Mast cells are required for angiogenesis and macroscopic expansion of Myc-induced pancreatic islet tumors. Nat Med 13:1211–1218

Spel L, Boelens JJ, Nierkens S, Boes M (2013) Antitumor immune responses mediated by dendritic cells: How signals derived from dying cancer cells drive antigen cross-presentation. Oncoimmunology 2:e26403

Spiegelberg HL (1990) Fc receptors for IgE and interleukin-4 induced IgE and IgG4 secretion. J Invest Dermatol 94:49S–52S

Spillner E, Plum M, Blank S, Miehe M, Singer J, Braren I (2012) Recombinant IgE antibody engineering to target EGFR. Cancer Immunol Immunother 61:1565–1573

Stone KD, Prussin C, Metcalfe DD (2010) IgE, mast cells, basophils, and eosinophils. J Allergy Clin Immunol 125:S73–s80

Tan SY, Fan Y, Luo HS, Shen ZX, Guo Y, Zhao LJ (2005) Prognostic significance of cell infiltrations of immunosurveillance in colorectal cancer. World J Gastroenterol 11:1210–1214

Tang Y, Lou J, Alpaugh RK, Robinson MK, Marks JD, Weiner LM (2007) Regulation of antibody-dependent cellular cytotoxicity by IgG intrinsic and apparent affinity for target antigen. J Immunol 179:2815–2823

Teng MW, Kershaw MH, Jackson JT, Smyth MJ, Darcy PK (2006) Adoptive transfer of chimeric FcepsilonRI gene-modified human T cells for cancer immunotherapy. Hum Gene Ther 17:1134–1143

Teo PZ, Utz PJ, Mollick JA (2012) Using the allergic immune system to target cancer: activity of IgE antibodies specific for human CD20 and MUC1. Cancer Immunol Immunother 61:2295–2309

Turner MC, Chen Y, Krewski D, Ghadirian P (2006) An overview of the association between allergy and cancer. Int J Cancer 118:3124–3132

Untersmayr E, Scholl I, Swoboda I, Beil WJ, Forster-Waldl E, Walter F, Riemer A, Kraml G, Kinaciyan T, Spitzauer S, Boltz-Nitulescu G, Scheiner O, Jensen-Jarolim E (2003) Antacid medication inhibits digestion of dietary proteins and causes food allergy: a fish allergy model in BALB/c mice. J Allergy Clin Immunol 112:616–623

Veggian R, Fasolato S, Menard S, Minucci D, Pizzetti P, Regazzoni M, Tagliabue E, Colnaghi MI (1989) Immunohistochemical reactivity of a monoclonal antibody prepared against human ovarian carcinoma on normal and pathological female genital tissues. Tumori 75:510–513

Vogel CL, Cobleigh MA, Tripathy D, Gutheil JC, Harris LN, Fehrenbacher L, Slamon DJ, Murphy M, Novotny WF, Burchmore M, Shak S, Stewart SJ (2001) First-line Herceptin monotherapy in metastatic breast cancer. Oncology 61(2):37–42

von Wasielewski R, Seth S, Franklin J, Fischer R, Hubner K, Hansmann ML, Diehl V, Georgii A (2000) Tissue eosinophilia correlates strongly with poor prognosis in nodular sclerosing Hodgkin's disease, allowing for known prognostic factors. Blood 95:1207–1213

Wan T, Beavil RL, Fabiane SM, Beavil AJ, Sohi MK, Keown M, Young RJ, Henry AJ, Owens RJ, Gould HJ, Sutton BJ (2002) The crystal structure of IgE Fc reveals an asymmetrically bent conformation. Nat Immunol 3:681–686

Wang SY, Racila E, Taylor RP, Weiner GJ (2008) NK-cell activation and antibody-dependent cellular cytotoxicity induced by rituximab-coated target cells is inhibited by the C3b component of complement. Blood 111:1456–1463

Wang SY, Veeramani S, Racila E, Cagley J, Fritzinger DC, Vogel CW, St John W, Weiner GJ (2009) Depletion of the C3 component of complement enhances the ability of rituximab-coated target cells to activate human NK cells and improves the efficacy of monoclonal antibody therapy in an in vivo model. Blood 114:5322–5330

Wasiuk A, de Vries VC, Nowak EC, Noelle RJ (2010) Mast cells in allergy and tumor disease. In: Penichet ML, Jensen-Jarolim E (eds) Cancer and IgE: introducing the concept of allergooncology. Springer, New York, pp 137–158

Watanabe N, Bruschi F, Korenaga M (2005) IgE: a question of protective immunity in *Trichinella spiralis* infection. Trends Parasitol 21:175–178

Weichman BM, Hostelley LS, Bostick SP, Muccitelli RM, Krell RD, Gleason JG (1982) Regulation of the synthesis and release of slow-reacting substance of anaphylaxis from sensitized monkey lung. J Pharmacol Exp Ther 221:295–302

Weiner LM, Surana R, Wang S (2010) Monoclonal antibodies: versatile platforms for cancer immunotherapy. Nat Rev Immunol 10:317–327

Weiner LM, Murray JC, Shuptrine CW (2012) Antibody-based immunotherapy of cancer. Cell 148:1081–1084

Weng WK, Levy R (2001) Expression of complement inhibitors CD46, CD55, and CD59 on tumor cells does not predict clinical outcome after rituximab treatment in follicular non-Hodgkin lymphoma. Blood 98:1352–1357

Winau F, Winau R (2002) Emil von Behring and serum therapy. Microbes Infect 4:185–188

Winau F, Westphal O, Winau R (2004) Paul Ehrlich–in search of the magic bullet. Microbes Infect 6:786–789

Ying S, Robinson DS, Meng Q, Barata LT, McEuen AR, Buckley MG, Walls AF, Askenase PW, Kay AB (1999) C–C chemokines in allergen-induced late-phase cutaneous responses in atopic subjects: association of eotaxin with early 6-hour eosinophils, and of eotaxin-2 and monocyte chemoattractant protein-4 with the later 24-hour tissue eosinophilia, and relationship to basophils and other C–C chemokines (monocyte chemoattractant protein-3 and RANTES). J Immunol 163:3976–3984

Yokota A, Yukawa K, Yamamoto A, Sugiyama K, Suemura M, Tashiro Y, Kishimoto T, Kikutani H (1992) Two forms of the low-affinity Fc receptor for IgE differentially mediate endocytosis and phagocytosis: identification of the critical cytoplasmic domains. Proc Natl Acad Sci USA 89:5030–5034

Zhang K, Kepley CL, Terada T, Zhu D, Perez H, Saxon A (2004) Inhibition of allergen-specific IgE reactivity by a human Ig Fcgamma-Fcepsilon bifunctional fusion protein. J Allergy Clin Immunol 114:321–327

Zhang QW, Liu L, Gong CY, Shi HS, Zeng YH, Wang XZ, Zhao YW, Wei YQ (2012) Prognostic significance of tumor-associated macrophages in solid tumor: a meta-analysis of the literature. PLoS ONE 7:e50946

Zhu D, Kepley CL, Zhang M, Zhang K, Saxon A (2002) A novel human immunoglobulin Fc gamma Fc epsilon bifunctional fusion protein inhibits Fc epsilon RI-mediated degranulation. Nat Med 8:518–521

Index

A
Affinity maturation, 9
AID, 24–27, 30, 31
Allergic disorders, 93
Allergic rhinitis, 6, 39, 46, 49, 50, 54, 96
Allergy, 6
Anaphylaxis, 1, 2, 8
Anti-IgE, 14, 39, 40, 42, 43, 45–56, 97
Asthma, 39, 42, 44, 47–50, 53, 56, 94–96, 117
Atopic Dermatitis, 2, 50, 53, 137

B
B cell memory, 11

C
Cancer immunotherapy, 111, 136, 138
Class switch recombination, 3, 21, 22, 24, 28
Component-resolved diagnosis (CRD), 93, 102
Cow's milk allergy, 51
Cross-reactive carbohydrate determinants (CCDs), 98
Cross-reactivity, 93

D
Degranulation, 2, 42, 46, 56, 66–68, 70–75, 77–82, 93, 94, 97, 104, 114–117, 124, 125, 129–131, 133, 138
Direct switching, 8, 12

E
Eosinophils, 41, 42, 47, 91, 92, 96, 112–115, 117, 118, 123, 124, 127–129, 134, 136
Extra-membrane proximal domain (EMPD), 12, 13

F
FcεRI, 40, 63, 65–76, 80–82
FcεRII, 112–114, 136, 137
Food allergy, 50, 51, 54, 137

G
Germinal center, 1, 3, 6, 9
Germline transcription, 3, 7, 24, 25
G-protein-coupled receptors, 76

H
Helminths, 92
High-affinity antigen, 73

I
I epsilon exon, 23, 25
IgE cross-reactivity, 99
IgE, high affinity, 1, 3, 8, 9, 12, 14, 32, 41, 66, 97, 104
IgE immunotherapy, 109, 119, 120, 135–138
IgE, low affinity, 8, 9, 31, 32, 42
IgE, membrane form, 11, 129
IgE memory, 1, 3, 11, 12, 14
IgE receptor, 41
IgE, secreted form, 13
Immunoglobulin G1 (IgG1), 5–12, 25, 26, 28–32, 45, 111–113, 115, 119, 122, 123–125, 127, 129–132
Immunoglobulin E (IgE), 40, 63, 65–73, 75, 76, 81, 82, 92, 93
Interleukin 4 (IL-4), 3–7, 12, 28–32, 71, 79, 113, 114

L
LAT1, 66, 74, 75
LAT2, 74, 75
Low-affinity antigen, 73

M
Mast cells, 66
Microarray biochips, 103

O
Omalizumab, 45

P
Peanut allergy, 98, 102
Peptide cross-reactivity, 100
Plasma cells, 1, 3, 10
Polyadenylation signal, 12

R
Rush Immunotherapy, 50

S
Sequential switching, 3, 6, 8, 9, 11, 12, 14
Somatic hypermutation, 22, 24
Switch region, 23

T
T follicular helper (Tfh) cells, 5, 6, 9
Th2 cells, 5, 6, 14, 91, 117
TLR, 76, 81, 82
Transcription factors, 22, 23, 29

V
V(D)J recombination, 21–23

Printed by Printforce, the Netherlands